The Japanese and Indian Space Programmes: Two Roads Into Space

Springer
London
Berlin
Heidelberg
New York
Barcelona
Hong Kong
Milan
Paris
Santa Clara
Singapore
Tokyo

Brian Harvey

The Japanese and Indian Space Programmes: Two Roads Into Space

Springer

Published in association with
Praxis Publishing
Chichester

PRAXIS

Brian Harvey, M.A., H.D.E., F.B.I.S.
2 Rathdown Crescent
Terenure
Dublin 6W
Ireland

SPRINGER–PRAXIS BOOKS IN ASTRONOMY AND SPACE SCIENCES
SUBJECT *ADVISORY EDITOR*: John Mason, B.Sc., Ph.D.

ISBN 1-85233-199-2 Springer-Verlag Berlin Heidelberg New York

British Library Cataloguing in Publication Data
   Harvey, Brian, 1953-
      The Japanese and Indian space programmes: two roads into
         space. – (Springer–Praxis series in astronomy and space
         sciences)
      1.Astronautics – Japan   2.Astronautics – India
      I.Title
      629.4'0952

      ISBN 1-85233-199-2

Library of Congress Cataloging-in-Publication Data
   Harvey, Brian, 1953-
      The Japanese and Indian space programmes : two roads into space / Brian Harvey.
         p. cm. -- (Springer-Praxis books in astronomy and space sciences)
      Includes bibliographical references and index.
      ISBN 1-85233-199-2 (alk. paper)
       1. Astronautics--Japan. 2. Astronautics--India. I. Title. II. Series.

   TL789.8.J3 H37 2000
   387.8'0952--dc21   00-021284

„ Praxis Publishing Ltd, Chichester, UK, 2000
Printed by MPG Books Ltd, Bodmin, Cornwall, UK

Cover design: Jim Wilkie
Typesetting: Heather FitzGibbon, Christchurch, Dorset, UK

Printed on paper supplied by Precision Publishing Papers Ltd, UK

Table of contents

List of photographs, tables, figures and maps

PHOTOGRAPHS

TABLES

FIGURES

MAPS

Author's acknowledgements

The author wishes to thank all those who kindly assisted him with the provision of information, advice and photographs for this book. In particular, he would like to thank:

Japan
Professor Yaunori Matogawa, Director, Kagoshima Space Centre
Dr Chiaki Mukai, NASDA
Keiichi Nagamatsu, Director, Industrial Affairs Bureau, Keidanren, Tokyo
Maki Sato, Japan Satellite Systems Inc, Tokyo
Dr Nobihiro Tanatsugu, ISAS, Tokyo
Kanako Toshioka, Institute of Space & Astronautical Science (ISAS), Kanagawa
Ryoko Umetsu, Administration Department, Aerospace Division, Nissan Motor

I especially wish to acknowledge the assistance of Professor Makoto Nagatomo, ISAS, and Dr Ryojiro Akiba, ISAS, who provided original material and translations concerning the life of Dr Hideo Itokawa.

India
S.K. Bhan, National Remote Sensing Agency, Hyderabad
Dr George Joseph, ISRO Space Applications Centre, Ahmedabad
S.M.A.K. Khan, Head, Liquid Propulsion Systems Centre, Valiamala, Trivandrum
S. Krishnamurthy, ISRO, Bangalore
M.Y.S. Prasad, INSAT Master Control Facility, Hassan
D.P. Rao, National Remote Sensing Agency, Hyderabad
Professor U.R. Rao, former director, ISRO
S. Srinivasan, Vikram Sarabhai Space Centre, Trivandrum

Europe
Philip S. Clark, Molniya Space Consultancy, London
Rex Hall, London
Anders Hansson, London
NASDA, Paris

United States
Jim Harford, Princeton, New Jersey

For photographs, I am most grateful to the following: Institute of Space and Astronautical Science, National Space Development Agency, Nissan, Rocket Systems Corporation, Ryojiro Akiba, Indian Space Research Organization, Liquid Propulsion Systems Centre

Author's introduction

The early years of space development have been well chronicled—from the sudden appearance of the Soviet Union as a global space power to the American adventure to send the first men to the Moon. Barely noticed in this epic human drama was the entry of a second group of countries into the space race—France in 1965, followed by Europe in 1979 and even small nations like Israel in 1988. What was perhaps most remarkable about the second wave of spacefaring nations was that three of them were in Asia—Japan launched its first satellite in February 1970 and China only two months later. After several failures, India launched its first satellite ten years later in 1980.

Asia is now the main global growth area for rocketry and satellite technology. Although there have been recent setbacks to the economies of the region, strong economic growth and a vision of space exploration in the service of development have driven the rapid expansion of the space industry there. Governments see the use of advanced technologies as an essential cutting edge of economic, industrial and rural development. Tens of thousands of people are now employed in space-related industries.

Three countries are now striving for leadership of the Asian space industry: China, Japan and India. China developed communications satellites and recoverable capsules, and is now preparing to launch its first cosmonauts into space. The history of the Chinese space programme has already been published. Now is the time to examine the other two prime Asian space powers Japan and India.

Japan was the fourth nation to send satellites into Earth orbit. The country has developed a series of launchers, run a consistent scientific programme of small satellites and sent space probes to Comet Halley, the Moon and Mars. The Indians have developed their own launchers, mastered the techniques of solid rocketry and put their satellites at the disposal of rural education, communications and the development of the country's natural resources in a series of model programmes. Both countries have followed different roads into space; both merit detailed analysis.

Japan has carried out more than fifty launchings of its own rockets and satellites; ten of its satellites have been put into orbit by other countries. Japan is:

- the fourth country into space, after the USSR, USA and France;
- the third country to send spacecraft to geostationary orbit;
- the third country to send spacecraft to the Moon and Mars (after the USSR and USA);
- an important participant in the International Space Station, with its own orbiting laboratory, *Kibo*.

Five Japanese have flown in space—four on the American space shuttle (one made a space walk); and a Japanese journalist was the first media representative into space when he flew to the *Mir* space station in 1990.

India's space programme is smaller than that of Japan, but is equally interesting, demonstrating how space technology may be used to advantage by a developing nation. India has launched nine of its own satellites; 16 Indian satellites have been launched abroad (by Europe, the USSR/Russia and the United States). India has put its space programme to work for telecommunications, weather forecasting and remote sensing in a determined effort to promote education, rural development and sustainable economic growth. Its Earth resources satellites are the most advanced in the world. Both countries had towering father figures who inspired their space development: Hideo Itokawa in Japan and Vikram Sarabhai in India. This is their story.

NOTE ON TERMINOLOGY

The nomenclature of both Indian and Japanese spacecraft presents a number of problems. This book uses the formats likely to be most familiar to existing observers of the two programmes. On the Japanese side, following the prevailing custom, the main NASDA rockets have been given Roman numeral identifications (N-I, N-II, H-I, H-II) whereas the ISAS launchers have been given Arabic number identifications (*Mu-3*, *Mu-5*). Japanese satellites are referred to both by their technical designations (e.g. Engineering Test Satellite 6) and by their Japanese name given to them once they enter orbit (*Kiku 6*).

Part I
Japan

1

Origins

On 16 April 1944, the low, slender silhouette made its way slowly out of the German submarine base of Lorient in occupied France and headed out into the deep waters of the Atlantic. Nothing unusual, for this scene had taken place hundreds of times over the previous three years. Except that the submarine was not a German U-boat, but the Imperial Japanese Navy submarine *I-Go-29* with Commander Eiichi Iwaya on board. It was heading for Japan with the German Reich's most secret rocket engine designs. Soon after, another Japanese submarine, under the command of Haruo Yoshikawa left Lorient with more, priceless, rocket designs.

I-Go-29 reached the Japanese naval base of Singapore on 14 July where Cdr Eiichi Iwaya disembarked, bringing with him as many blueprints as he could carry, and took a plane to Japan. It was as well he did, for *I-Go-29* hit a mine and sank off Taiwan on 26 July. The second submarine was caught by the allied air forces and sunk, and Cdr Yoshikawa was drowned.

Japan had come to learn of German rocket advances the previous year. Under the Japan–Germany Technical Exchange Agreement, 1943, the two countries had agreed to share technical information. Japan had become aware of American plans to bomb Japan with a long-range high-altitude bomber, the B-29, and asked Germany for help. In response, the German Air Force, the Luftwaffe, had told the resident senior Japanese naval office resident in Berlin, Cdr Eiichi Iwaya, in March 1944, that Germany was developing a rocket-propelled fighter able to climb to 10 000 m in 3.5 minutes, the Messerschmitt 163, or *Komet*. Germany had high hopes that the *Komet* could reach the marauding American and British bombers and use its extraordinary speed to cause havoc to their formations. Early the following month, on 6 April, Cdr Iwaya saw the Me-163 on test at Augsburg and was astonished by its small size, stubby delta wings and vertical ascent close to the speed of sound. The Germans agreed to hand over the blueprints of the Me-163 and its HWk 509A rocket engine, a manual on how to handle its tricky fuel (called T-stoff) and other information on rocket propellants.

JAPAN'S ROCKET PLANE

Back in Japan, on 20 July 1944, Mitsubishi's Nagoya plant was ordered to go ahead with the manufacture of a rocket frame and engine. The project was called *Shusui*, but given the navy designation of J8M-1 and the army name of Ki-200, the engine being called the Tokuro-2. *Shusui* was to climb to over 12 000 m under rocket power and attack the B-29 with its two 30-mm machine guns its pilot would then bring it back in a gliding descent to it base and land on a skid. The fuel consisted of two tonnes of 80% hydrogen peroxide and 20% hydrazine, methanol and copper potassium cyanide.

As the allies drew closer to Japan, getting the rocket ready was a race against time. The Japanese were not helped by the loss of the two submarines, for they contained more extensive documents and samples which could not have been carried on Cdr Iwaya's plane. Good progress was made with the body of the plane, the first wooden models being built by August 1944. A glider version called *Akigusa* ('autumn grass' in Japanese) was built and flown by Lt Toyohiro Inuzuka at Ibaraki.

The engine proved more problematical. No sooner was the Tokuro-2 tested that November than the factory was hit by the Mikawashima earthquake and then a B-29 raid. The site had to be moved to Natsushima. The engine was not completed until 2 July 1945. On 7 July 1945, Lt Inuzuka brought the *Shusui* up to 250 m, using 16 seconds of fuel for an initial test. However, while gliding back, the wingtip hit an observation tower and landed hard. Although there was no fire, Lt Inuzuka was badly injured and died early the following morning. Then the fuelling problems which had dogged the German development of the Me-163 came to afflict the Japanese. In what was to have been the first operational test of *Shusui*, on 15 July 1945, the engine exploded and killed the pilot, Lt Skoda.

The *Shusui* was almost identical to the Me-163. The Japanese army and navy ordered 155 of them and work began on setting up a *Shusui* production base at Atsugi. A *Shusui* corps was formed, the 312 Naval Flying Corps, based at Yokosuka. In the event only four flight models were made. The third test of *Shusui* was set for August 1945. It was delayed and, by the time it was rescheduled, Japan had surrendered.

INTRODUCING HIDEO ITOKAWA

Although *Shusui* was the high point of Japan' wartime rocket effort, in fact Japan's use of rockets dated back to the 1920s. The Japanese space programme owes its origins to Professor Hideo Itokawa, a teacher at the Institute of Industrial Science at Tokyo University, later the Institute of Space and Aeronautical Science.

Hideo Itokawa was born 20 July 1912. The name Hi-De-O means, in literal Japanese, 'fire coming out of man'. The young boy earnestly set about living up to his title. With his older brother, in his last year at Azabu Nanzan primary school, he constructed electric cannon ignited by alcohol and shot out of glass tubes. He competed with his brother to see whose pellets would fire the furthest, remarkably never injuring himself in the process, to the surprise and relief of his mother. He left primary school two years early to made a precocious entry to secondary school. By the age of 12 he was pondering the problems of electromagnetic propulsion, devoting his daylight hours and apparently some sleepless nights to overcoming the challenge. He persuaded a local blacksmith to make some metal coils for him, though he later commented that it was difficult to get the right materials in Japan in the 1920s. Reaching the newly opened First Tokyo City Middle

School, he carried out electromagnetic experiments, prevailing upon the physics teacher, Genji Kawashima, to give him the run of the school's science laboratory. Most of the experiments he made there were built from first principles, since, as he said later, the right theoretical books simply were not available. Itokawa then entered the school of aeronautical science in the University of Tokyo, one of the first such university schools in the world.

When the war came, Hideo Itokawa was a designer for the Imperial Japanese Army and was an engineer for the Nakajima Aircraft Company from 1939 to 1945, though he continued his research and teaching work in the University. Although aircraft-makers such as Mitsubishi were better known to their American and British opponents, Nakajima was one of imperial Japan's main wartime planemakers. Nakajima made the *Gekko* and *Saiun* fast reconnaissance plane, the Ki-49 which bombed Darwin, the *Tenyan* torpedo bomber, the outstanding Ki-84 *Hayate* fighter and, toward the very end, the Ki-115 special attack (*kamikaze*) fighter. Itokawa led the team which developed the Ki-43 *Hayabusa* 'Oscar" attack fighter, a highly manoeuvrable plane used by the Japanese army throughout the Pacific. In addition to *Shusui*, Japan developed a rocket-propelled kamikaze plane, the *Okha*, or cherry blossom. The *Okha* was carried to its target by a mother plane. Once it was dropped, the pilot used its rocket engines to speed the plane, with its two-tonne warhead, into a 800 km/h dive onto its target.

The *Shusui* and *Okha* were the best-known of Japan's wartime rockets, but not the only ones. From 1942, short-range 700-kg rockets, 50 cm in diameter were used to supplement the firepower of the army's artillery. As American air raids began to devastate the mainland, rockets were designed by the Naval Technical Assistance Unit to attack the bombers. The unit's first rocket was a 25-kg surface-to-air rocket. The unit developed, in the Yokosuka dockyard, the *Funryu* anti-ship solid-fuelled missile (*Funryu* 1 and 2) and a more ambitious anti-ship liquid-fuelled guided missile (*Funryu* 3 and 4), flying one to an altitude of 32 km. Kawasaki and Mitsubishi between them developed air-to-surface missiles with a thrust of 250 kg for 75 seconds and a range of 8 km. Japan's wartime rockets would have played a more significant role had they reached mass production.

After the war, Japan was not permitted to build either aircraft or rockets, so Itokawa went to work in the medical school in the University of Tokyo, concentrating on acoustics (he made an improved model of violin). His wartime experience had taught him something of the immense propulsive force of the rocket and he was aware of the development of the *Shusui*. In the years which followed the war, the United States and the Soviet Union sent re-engineered German V-2 rockets high into the atmosphere with small scientific payloads, cameras and even animals. He visited the United States in the early 1950s and was enormously impressed with the level and pace of American rocket development.

Itokawa hoped that these developments could be matched in Japan. The legal restriction on aeroplane and rocket development was lifted in 1951 under the San Francisco peace treaty. With his colleagues and students, he began to design and build small solid-fuel rockets under the aegis of the Institute of Industrial Science at the University of Tokyo, where he and his colleagues formed what was called the Avionics and Supersonic Aerodynamics Research Group (AVSA). This group took in a mixture of ex-wartime designers, civil engineers, architects and experts in mechanics and physics. Their first studies were theoretical. At the second meeting of the group in April 1954, discussion centred on a feasibility study by Hideo Itokawa on a rocket transport plane able to fly the Pacific in 20 minutes.

However, it was ultimately more important for the group to undertake some practical activity. The AVSA was able to obtain government grants of ¥3.3m (€29 464) from the ministries of education and international trade, supplemented by Fuji Precision Co. to pursue their research. Unusually for the post-war pioneers, Itokawa chose to go down the road of solid-fuel rockets. The German V-2 and most of the post-war American and Russian rockets were liquid-fuelled. These rockets had two fuel tanks, one containing fuel, the other oxidizer, which were pumped to great pressure and then fed into a combustion chamber for ignition and burning. Solid-fuel rockets, by contrast, were more akin to the traditional firework, containing a single tank filled with a grey, sludge-like substance which was poured in and later solidified. Solid fuel rockets were less sophisticated (there was no plumbing involved) and offered potentially greater thrust, but were harder to control and could not be turned off or throttled. In fact, Itokawa and his colleagues had such limited resources that solid fuels were the only realistic possibility open to them.

THE FIRST ROCKETS

The first such rockets were really tiny. The first was—quite appropriately—called *Pencil*, being just 23 cm long and 1.8 cm in diameter. Using relatively cheap raw materials, Itokawa made 150 firings of *Pencil* and was able to reach conclusions about the best type of fuel, the most suitable configuration of nozzle and the shape of stabilizing fins. They were fired horizontally at first, at Kokubungi, Tokyo. By April 1955, Itokawa and his colleagues felt sufficiently confident in their work to give a public demonstration of *Pencil* in a Tokyo suburb. In the end, 29 were launched. Public reaction to the experiments was divided: most people enjoyed the experiments, but some scientists criticized them as silly and meaningless.

However, the environs of Tokyo would not long remain a safe launch base, so in March 1955 Itokawa and his university colleagues established a beach launch site at Michikawa, on the Sea of Japan in Akita and on 6 August 1955 launched a *Pencil 300* at the new site. There, in what was now called the Akita range, they went on to develop the next stage after *Pencil*. The new rocket, in deference to its diminutive size, was called *Baby*, and was first launched from Akita in late August 1955. *Baby* had a diameter of 8 cm and held a full kilogram of propellants—mainly gunpowder. Thirty-six *Babys* were fired that year and one version reach 6 km.

Things were very primitive in those days. None of those involved in the project had a car, so rocket equipment was brought in by horse! Pictures show Dr Hideo Itokawa counting down a rocket under a wooden hut as crowds of onlookers gathered on a nearby hillside. The control panel comprised some large electric bulbs. After a few years, Itokawa and his colleagues were able to construct three small concrete buildings and some temporary wooden ones. The rockets blew up often enough and Itokawa tried to laugh away the problems as inevitable in such an esoteric science.

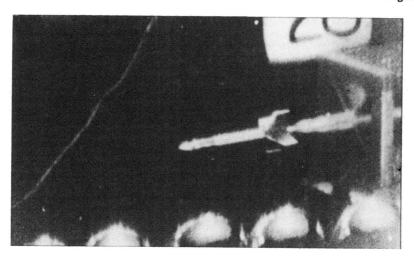

Pencil rocket.

Neither Itokawa nor his colleagues were sure of the best way forward in rocket development at that time. Following the end of the *Pencil* programme, Itokawa asked his colleague Ryojiro Akiba to give some attention to the problem of reaching ever-higher altitudes. Akiba was able to borrow a mechanical analogue computer from an electronics professor and do some of the basic calculations. He suggested multistage rockets and the use of rockoons.

The theory of the rockoon is that a balloon carry a ready-to-go rocket to a high altitude, at which point its engine is lit to send it into space. The balloon, using passive energy, saves the rocket the energy required for the earliest and most difficult part of its ascent through the atmosphere. The Science Council of Japan approved the rockoon project in spring 1956, though the money in fact came from the Yomiuri newspaper company and the project was to be implemented by the Japanese Rocket Society. Two rockoons were successfully launched in 1961 from Rokkasho, Aomori, but the project ended at that stage. The rockoon idea was to be resurrected for shuttle tests many years later.

In the event, multistage rockets offered more immediate promise to the altitude problem. Itokawa built a two-stage version of *Baby*, a rocket 150 cm long, one which lit a second stage when the first one had exhausted its propellant. In the top part of this *Baby*, Itokawa installed a tiny radio transmitter and in another version a small camera and parachute.

Itokawa campaigned for the Japanese government to participate in the International Geophysical Year (IGY), designated for 1957–58. As far back as 1954, Japanese scientists Kenichi Maeda and Takeshi Nagata had pressed the Ministry of Education that Japan participate in the year. The ministry initially refused, but over the new year holidays an official in the ministry chanced upon an article called 'Rocket transport' in the newspaper *Mainichi*. He persuaded his departmental colleagues to change their minds

---

The great space designers

US	Robert Goddard, Maxime Faget
Austria	Hermann Oberth, Eugene Sänger, Irene Bredt
Germany	Wernher von Braun
China	Tsien Hsue-Shen
Russia	Konstantin Tsiolkovsky, Sergei Korolev, Yuri Kondratyuk, Friedrich Tsander, Vladimir Chelomei, Valentin Glushko, Semion Kosberg, Mikhail Yangel
Japan	Hideo Itokawa
India	Vikram Sarabhai

---

and in February AVSA was allocated a budget of ¥17.4m (€155 357) to enable their participation in the IGY.

AVSA was renamed the Sounding Rocket Group and began to fire more ambitious rockets in 1957. The new series was called the K programme, or kappa. A K-3 reached an altitude of 18 km in July 1957. They proceeded with the improved K-4, but trouble with the propellant mixtures led to two explosions in September 1957.

Although few realized this at the time, the IGY had sparked off the first phase of the space race, with the Soviet Union and the United States both racing to put scientific satellites into orbit before the end of 1957. Unnoticed in the excitement which marked the launching of the first Sputnik in October 1957, the Japanese government first gave money to its scientists to make a contribution to the IGY. Japan was one of nine observing locations for the year and the government authorized Itokawa to go ahead with the building of the a new *Kappa* sounding rocket.

SOUNDING ROCKETS

Japan's IGY sounding rocket was the *Kappa-6*, a two-stage solid-fuel sounding rocket 6.5 m long, 25 cm in diameter and weighing 260 kg, ten times larger than *Pencil*. At this stage, the university's rocket used letters from the Greek alphabet, *kappa* being later followed by *lambda* and then *mu*. As a sounding rocket, *Kappa* was a radical development, for until that time almost all sounding rockets designed to reach altitude had been liquid-fuelled. The university stuck with solid fuels, using reinforced plastic and aluminium for the nozzles and rocket bodies respectively. The *Kappa-6* was the first of its kind to have two stages, the upper rocket taking over when the first stage exhausted its propellants.

More important, the *Kappa-6* could carry 12 kg of scientific instruments. Thirteen *Kappa-6* rockets were launched during the IGY, reaching altitudes of up to 60 km—the beginnings of space—where they collected information on the upper atmosphere and cosmic rays. The programme continued long after the IGY ended.

Hideo Itokawa at launch site.

The first *Kappa-6* was launched on 16 June 1958. The second stage shut down early for no apparent reason, but a second launching three days later was successful. Thirteen *Kappa-6s* were launched as Japan's contribution to the IGY.

Noting their success, the government began to show increasing interest in rocketry. In 1958, the prime minister's office established a National Space Activities Council to discuss the best way forward in space research. Later, the government set up the National Space Development Centre of the Science & Technology Agency while the University of Tokyo's space activities were given new form as the Institute of Space and Aeronautical Science (ISAS) in 1959. Its first director was one of Itokawa's colleagues, Professor N. Takagi. ISAS in effect was a bringing together of the old Aeronautical Research Institute in the university and the rocket research scientists.

The *Kappa-6* became the basis for a series of new sounding rockets, each more impressive than its predecessor. The *Kappa-8*, weighing 1.5 tonnes, 11 m long, with a payload of 90 kg, could reach an altitude of 200 km, the altitude of a satellite in orbit. It was first flown in September 1959 and used steel motor case welding techniques of the type developed by the shipbuilding industry. The *Kappa-9L* was the first three-stage sounding rocket. In April 1961, the month Yuri Gagarin flew around the world in orbit, the *Kappa-9L* soared to 310 km above the Earth. Its successor, the long-lasting *Kappa-9M*, reached the same height, but with four times greater a payload. The last of the series, the *Kappa-10*, launched late 1965, reached 700 km. *Kappa* rockets were later exported to Yugoslavia and Indonesia.

Japan's next rocket was the *Lambda*. Although originally built as a high-performance sounding rocket, it later became the rocket that first gave Japan access to Earth orbit. The purpose of *Lambda* was to reach altitudes of 3000 km, well into space. A testing centre for solid and liquid fuel rockets was established by the university at Noshiro.

KAGOSHIMA LAUNCH SITE

Akita range was no longer suitable for such a more sophisticated space programme, so the existing projects were relocated. In early 1961, Professor Itokawa began to search for a new and more suitable launch site, away from populated areas. The site chosen was Kagoshima, on the southern edge of Japan, where work started in February 1962. At the tip of the Japanese island of Kyushu, the site was relatively uninhabited and enabled launchings to be safely made across the entire Pacific ocean. They moved not a moment too soon, for in May 1962 there was a dramatic accident to a *Kappa* sounding rocket at Akita: the first stage exploded and the second stage, still live, skipped across the sea and headed straight toward a cottage where it burned out. No one was injured, but everyone got a fright.

Kagoshima had two drawbacks, however. The first was that the site was mountainous, so facilities had to be cut out of the neighbouring hills with heavy earth-moving equipment; and the second was the fishing problem. Fishing is a vitally important industry in Japan, and indeed the Japanese are reputed to eat more fish per head of population than any other nation on Earth. Local fishermen objected strongly to the noise of ascending rockets which, they argued, upset the local fish. In any case, range safety officers insisted the area downrange of a launch pad be cleared while a rocket was being fired, lest débris from an explosion fall on fishing boats. The strap-ons for the later H-I

Kagoshima launch centre.

rocket, for example, crash some 25 km downrange, right in the middle of the fishing zone.

The upshot of the stand-off was a bizarre compromise whereby rockets could only be launched in two limited periods of 90 days altogether during the year, based on February and September. Whilst a fair compromise on paper, not even the most efficient space industries have ever managed to organize their space launches to fit perfectly into such a schedule; nor indeed are the alignments of the Moon and planets always coincidental with the Japanese fishing seasons.

Clearance of the 510 ha site took two years, but by the time it was completed for its official opening in December 1963, Japan had a full launch facility with *Kappa* and *Lambda* pads, control, radar and tracking facilities. The first *Lambda* was launched from Kagoshima in July 1964. This was the *Lambda-3*, a three-stage rocket, 19 m long, 73.5 cm in diameter, weighing 7 tonnes. Its first flight lasted 17 min, in the course of which it flew 1000 km high and impacted 1090 km downrange in the Pacific. The immediate purpose of the *Lambda* launchings was to enable Japanese participation in the next international year, which was the International Year of the Quiet Sun, marking the minimum levels of solar activity. During the solar minimum, Japan sent up 34 *Lambda* and *Kappa* sounding rockets. The third *Lambda* was used to study the radiation of our own Milky Way galaxy and was the first Japanese rocket to be launched by night. Public interest began to grow.

REACHING EARTH ORBIT

The *Lambda-3* marked the limits of what could be achieved by sounding rockets. By the mid-1960s, it was logical and becoming ever more possible for Japan to proceed to the next step, which was to put a satellite into orbit. In 1960, Hideo Itokawa and Ryojiro Akiba co-authored a paper in which they outlined how a small satellite could be put into orbit by adding a motor to the upper stage of what was in effect a large sounding rocket. In 1962, Itokawa and his colleagues presented a report entitled 'Tentative plan for a satellite launcher'. The Science Council of Japan held a symposium on a scientific satellite the following year, one which addressed the following questions:

• Can Japan still make contribution to science, despite its late start?
• Is a satellite project feasible?
• Can Japan achieve a satellite using indigenous technology, or should it rely on the United States?
• How can a satellite be tracked without a tracking centre outside Japan?

Events began to move at a faster pace now. In 1965, the National Space Activities Council gave the go-ahead for a scientific satellite programme proposed by ISAS. Professor Itokawa proposed the development of a new rocket, the *Mu*, as an operational satellite launcher and this was authorized in August 1966. Professor D. Mori was appointed project director.

In the meantime, a new version of the *Lambda* was developed, the *Lambda-4S*, under the direction of Professor T. Nomura. Itokawa had the notion that this launcher might be capable to getting a satellite into orbit before the *Mu* was ready, at lower cost and much

First successful orbital flight.

Lambda launches	
26 Sep 1966	Failure
20 Dec 1966	Failure
13 Apr 1967	Failure
22 Sep 1969	Failure
11 Feb 1970	*Ohsumi*

sooner. The first stage had a thrust of 36 970 kg, augmented by two 13 150 kg thrust solid strap-on boosters. The second stage had 11 800 kg thrust, the third 6580 kg thrust, and the fourth 816 kg. *Lambda* was a small, pencil-shaped launcher, with fins at the bottom and middle. Weighing 9 tonnes, with a thrust of 53.5 tonnes, the *Lambda-4S* was 16.9 m tall. The cost of development was ¥118m (€1.05m). It was as well they concentrated on the *Lambda*, for the *Mu* project was abandoned as too expensive (though a much later rocket was to bear the same name).

The difference from its predecessors was that the *Lambda-4S* had a small fourth stage comprising a small solid-rocket motor with a tiny capsule with instruments attached. In essence, the launching technique was to fire an unguided sub-orbital rocket but to use a small motor at the peak of the flight to kick a payload into orbit, a technique since used by many of the smaller space nations. In size, the *Lambda* was the smallest rocket ever used to get a spacecraft into orbit. Though taller than the later British *Black Arrow*, it was much thinner.

A precursor version of the *Lambda-4S* was tested in ballistic flight in March 1966, but it failed when the vanes of the third stage failed to control the spin of the rocket. A second test that summer worked, the version being called the *Lambda-3H*. Now the scientists felt they could aim at orbital flight.

Japan's first attempt to orbit a satellite was made on 16 September 1966 when the pyrotechnic devices on the last stage failed to fire. More heart-breaking failures followed. On the second attempt, the despin motor failed and the last stage separated prematurely. On the third attempt, the third-stage motor did not fire. Japanese newspapers criticized Itokawa's leadership, forcing him to resign from ISAS in April 1967.

Working on without him, Japanese scientists did not meet with success until the orbiting of the *Ohsumi* satellite on 11 February 1970 by the *Lambda-4S* rocket from the Kagoshima Space Centre, *Ohsumi* being named after the peninsula where the space

First satellites in orbit

Oct 1957	Soviet Union (*Sputnik 1*)	84 kg
Jan 1958	United States (*Explorer 1*)	14 kg
Nov 1965	France (*Astérix*)	40 kg
Feb 1970	Japan (*Ohsumi*)	38 kg
Apr 1970	China (*Dong Fang Hong 1*)	173 kg
Oct 1971	Britain (*Propero*)	66 kg
Jul 1980	India (*Rohini 1B*)	35 kg
Sep 1988	Israel (*Offeq 1*)	156 kg

Lambda-4S

Length: 16.52 m
Diameter: 73.5 m
Weight: 9.4 tonnes

Stage	Engines	Thrust, kg
Strap-on	Two SB-31 solid	13 150
1	735UP solid	36 970
2	735UP solid	11 800
3	500BP solid	6 580
4	480S solid	816

centre is located. Entering an orbit of 338 by 5150 km, inclination 31°, *Ohsumi* was a 12-kg instrument package with a beacon. The rocket was fired unguided out over the Pacific ocean, the stages dropping off one after the other. The third stage fell off at 85 km, by which the rocket had reached a speed of 5 km/sec. As the rocket reached the height of its trajectory, the fourth stage solid rocket motor fired, just enough to give the small satellite the final kick into orbit. Hideo Itokawa was in a desert in the Middle East when his Arab driver shouted to him the good news and recalled later how he sobbed endlessly with delight to know that his lifetime's dream had come true.

Ohsumi weighed 38 kg, including the solid rocket motor which enabled it to reach orbit. The instrumentation was limited to a radio transmitter, battery, thermometer and accelerometer. The signals failed after seven orbits, but Japan had become the fourth nation in space.

INTRODUCING THE *MU-4S*

Ohsumi was, to all intents and purposes a test satellite. It was too small to carry any useful scientific instrumentation. The next step was to put in orbit a scientific satellite. A similar approach was followed by the Chinese, who launched a demonstration satellite in 1970 (*Dong Fang Hong*) but their first true scientific satellite the following year, *Shi Jian 1*. To launch a scientific satellite with a useful payload required a more powerful launch vehicle, in effect an orbital rather than a sub-orbital vehicle. Development of an orbital launcher was begun in the institute as far back as April 1963. A new launch pad for more powerful rockets was constructed at Kagoshima in 1966 with an assembly room and satellite processing facility. The new rocket, called the *Mu-4S*, used one large set of tail fins for stabilization.

The *Mu-4S* weighed 43 tonnes and its objective was to place in orbit a small scientific satellite which would study solar radiation for a year. As launchers go, the *Mu* was one of the smallest in the world, being comparable in size to the French *Diamant* or the American *Scout*. *Mu* was larger than the *Lambda*, being 23.6 m high, with a 1.42 m diameter, and having a thrust of 80 tonnes. The white and red *Mu-4S* swung outward from the launch tower at an angle. Although the first *Mu-4S* climbed perfectly into the sky on 25 September 1970, the fourth stage let the designers down and the satellite crashed to destruction downrange in the Pacific ocean.

Success was achieved the following year. *Tansei* followed *Ohsumi* into orbit on 16 February 1971 and *Shinsei* on 28 September 1971. *Tansei*, weighing 62 kg, entered orbit of 990 by 1110 km, inclination 30°. *Tansei* means 'light blue' in Japanese, the colours of Tokyo University. *Tansei* was designed to study plasma waves and density, electron particle rays, geomagnetism and electromagnetic waves. It carried a transmitter which operated for a week, though the satellite's orbit should keep it in space for a thousand years. The main purpose of the launch was to demonstrate the *Mu-4S* rocket. For Itokawa, *Tansei* was the realization of his dream of Japan launching it own small scientific satellites using modest resources.

	Mu-4S	
	Length: 23.6 m	
	Diameter: 1.42 m	
	Weight: 43.6 tonnes	
Stage	Engines	Thrust, kg
Strap-on	Two SB-310 solid	79 200
1	M-10	73 400
2	M-20	28 500
3	M-30	12 400
4	M-40	1 900

DISCOVERING A NEW RADIATION BELT

Japan's third satellite was *Shinsei*, meaning 'new star' in Japanese. *Shinsei* was 1.2 m high, weighed 65 kg, carried solar panels and was designed to measure solar and cosmic radiation, the ionosphere and solar activity. Entering orbit of 869 by 1865 km, inclination 32°, *Shinsei* was expected to remain in orbit for 5000 years. *Shinsei's* achievement was to identify a new, small radiation belt around the Earth, adding to those found by the American *Explorer* spacecraft in 1958. The belt was located at low altitude near the equator and emitted a new type of radio wave.

Denpa was Japan's third scientific satellite, the fourth to reach orbit (19 August 1972). 75 kg *Denpa* entered a highly elliptical orbit of 245 by 6291 km, 31.03° where its instruments analysed the Earth's ionosphere, geomagnetic field, electrons and plasma. The second and third stages were unable to reach the intended height due to strong head winds and the fourth stage had to be fired earlier and longer than planned. Despite their best efforts, ground control found that the satellite had entered orbit 40% lower and three times higher than planned. The sun sensor subsequently failed and there was then a voltage fault in the encoder, so only fragmentary information was returned. This was disappointing.

SECOND VERSION—THE *MU-3C*

A problem with the early Japanese rockets, great though their achievements were, was that they were unguided. This meant that everything depended on the rocket being fired at the right angle at take-off and the rocket keeping strictly to this flight path. Deviations were difficult to correct and the final orbit of the satellite was hard to determine. Accordingly, a new version of the *Mu-4* was developed, one designed to ensure a more accurate insertion of satellites into orbit. The *Mu-3C* had similar dimensions to the *Mu-4S*, but the second stage had thrust vector control with additional control motors; and the third stage was equipped with radio guidance. This achieved the desired results when on 16 February 1974, the *Mu-3C* put *Tansei 2* into an orbit of 284–3233 km, 31.2°, one quite close to the

```
Mu-4S launches
25 Sep 1970        Failure
16 Feb 1971        Tansei 1
28 Sep 1971        Shinsei
19 Aug 1972        Denpa
```

orbit planned. *Tansei 2* was followed by *Taiyo* which studied solar ultraviolet and X-rays. CORSA also called *Hakucho*, was a 100-kg satellite launched on *Mu-3C* on 21 February 1979 to study X-rays. It was Japan's first X-ray astronomy satellite. It had eleven X-ray detectors to survey the sky along the galactic plane and to detect gamma ray bursts from neutron stars.

```
Mu-3C launches
16 Feb 1974        Tansei 2
24 Feb 1975        Taiyo
21 Feb 1979        Hakucho
```

Mu-3C

Length: 20.2 m
Diameter: 1.41 m
Weight: 43.6 tonnes

Stage	Engines	Thrust, kg
Strap-on	Eight SB-310	77 600
1	M-10	75 000
2	M-22-TVC	28 400
3	M-3A	5 800

By the late 1970s, Japan had realized Itokawa's dream of a country able to put small scientific satellites into orbit using small rockets developed with limited means. Despite setbacks and disappointments, eight small satellites had been put into orbit. A reliable solid rocket booster had been introduced. Itokawa lived long enough to see the growing maturity of the Japanese space programme. Hideo Itokawa does not fit the normal mould of the prototypical rocketeer. He took part in the Tukiji Shogekijo theatre and read the works of Shakespeare translated by Tubouchi, most of all liking *The Merchant of Venice*, *Hamlet* and *Henry V*. He learned English in his spare time. At school he was absorbed in music and wanted to be a composer. In the event, Lindbergh's flight across the Atlantic interested him in aeronautical science instead, but he later became a cellist, his favourite piece being Schubert's unfinished symphony. As a student during the great depression, he was involved in leftist politics. Although disadvantaged by illness and small size, he loved sport and outdoor activities—basketball, swimming and skiing. Poor health ruled him out of the officer corps of the imperial Japanese army. When after the war he could not develop aeronautics he worked on neurological and brain problems at the university while writing a doctoral thesis on the acoustic qualities of the flute.

Rocket science was just one of his many lifetime interests, one which built on his university and engineering experiences. He was portrayed as 'Dr Rocket' in the Japanese media in the 1950s and 1960s, but he had always promised to himself that he would move on from rocketry once the *Lambda* and *Mu* projects came to fruition. To prepare for this move, he set up what he called the Systems Research Laboratory in 1964. He disagreed with some of his colleagues about the future of rocket and scientific research but what forced his hand was what he felt was unfair criticism of himself and the *Lambda* programme by the newspaper *Asahi Shimbun*. He resigned from the university in March 1967, setting up Japan's first think-tank, the Systems Research Institute, that May. However, he still found the attraction of space research irresistible and advised the Indian government on its nascent space programme until 1971. He was acquainted with the Apollo rocket scientist Wernher von Braun, with whom he shared the same year of birth.

Hideo Itokawa got involved from 1968 to 1970 in a project to build huge undersea flying-saucer-shaped oil storage in 1bn litre tanks whose aim was to give Japan a strategic oil reserve (a foresightful project in the light of the subsequent oil crisis). Next, he moved on to a project for a nuclear-powered ship, the *Mutsu*. In his retirement he wrote books and became a popular philosopher, scientist, economist and educator. He was, reluctantly, prevailed upon to write an autobiography. He found it difficult to do so, describing his character as being that of a boar in the forest—an animal which dashes forward and never looks back. So he wrote a 'personal history', but not in a book form but as a poem. He lived on to old age—86 years—until 21 February 1999, when he died of a brain infection.

2

The making of a space programme

Even as Professor Itokawa was developing a small-scale programme of scientific space exploration, moves were afoot to develop a more ambitious space programme, one which reflected growing Japanese confidence in its economy, manufacturing and science. In May 1968, the Japanese government formed the Space Activities Commission, SAC, whose function was to propose policies, submit proposals to the prime minister and bring coherence to work in the field. During the 1970s and 1980s the Japanese developed a series of licensed rockets—the N-I, N-II and H-I—and began to launch engineering and communications satellites. In order to kick-start domestic satellite communications, American launchers were also used.

FORMATION OF NASDA

Pressure for a more aggressive space programme came from Japanese industry. A Space Activities Promotion Council was set up (10 June 1968), bringing together 89 companies engaged in space activities whose role was to present the needs of space-based companies to government, to make proposals for the development of space activities and to ensure better public understanding of space exploration. Specifically, industrial and commercial interests made an early pitch for the future development of the space industry, the Japan Broadcasting Corporation (semi-governmental) and the Ministry for Posts & Telecommunications arguing that it should urgently address the problem of poor-quality television reception in the Japanese islands. In 1973, business organizations began to campaign for long-term planning to develop a space industry and other forms of advanced technology. The Nikkeiren association proposed a 15-year ¥5.5bn (€49m) plan for a range of applications satellites, a powerful liquid-hydrogen-fuelled rocket and a space laboratory.

Up till the late 1960s, spending on space research had gone to a number of different organizations and groups, although the university's successes meant that it attracted the most public attention. The amounts of public money spent had been modest enough— ¥1.369bn (€122m) over the years 1954–70. These resources went to the university and a number of government departments and agencies, principally the government's own

Science and Technology Agency, which had begun work on the development of liquid-fuelled rockets.

The government attempted to settle these different roles, responsibilities and functions by the establishment of the National Space Development Agency, NASDA, set up on 1 October 1969 under law #50, 23 June 1969. NASDA was formed out of the National Space Development Centre of the Science & Technology Agency and the Radio Research Laboratory of the Ministry of Posts and Telecommunication. Its brief was to develop artificial satellites, plan space programmes, develop and launch rockets, track and control satellites and develop the necessary technologies, facilities and equipment for this purpose.

NASDA was specifically charged with responsibility for the development of launch vehicles; the promotion of technologies for remote sensing; and the promotion of space experiments. The idea was that NASDA would take charge of launch vehicle development, operations and the development of technology and applications satellites. Tokyo University, through the Institute of Space and Aeronautical Science, would retain responsibility for sounding rockets and scientific satellites. Although ISAS could continue to develop its own small solid-fuel rockets, the delineation of responsibilities between the two involved a limit being set on the size of rocket to be launched by ISAS—a diameter of 1.41 m.

In retrospect, the 1969 law institutionalized, rather than resolved the divided nature of the Japanese space programme. Thirty years later, the same division between the scientific programmes (ISAS) and technological and industrial development (NASDA) remained intact. While specialization reduced some duplication (ISAS indeed concentrated on scientific satellites and small, solid-fuel rockets), Japan in effect developed two different, parallel space programmes, a unique situation for a civilian space programme. Each had its own fleet of rockets, launch sites, mission control and tracking systems. There has been little discussion in Japan of the potential waste and duplication of effort, though some scientists have drawn attention to the manner in which discord between the different agencies has impeded the smooth progress of the programme [1]. Moreover, ISAS literature rarely mentions the work of NASDA and vice versa—indeed it is possible for a novice to read of the work of one without even being aware of the work of the other! Only in recent years have the two collaborated on joint programmes (for example the J-1 rocket). Indeed, the formation of a joint working group in 1996 caused such soul-searching and debate about how to reconcile the different cultures of ISAS and NASDA—yet neither had difficulty working with foreign partners! Having said this, despite the division of work between ISAS and NASDA, many NASDA scientists came from ISAS. Proposals were eventually made in 2000 for a merger.

This is not to say that there is not some coordination between the two. In 1978, the Space Activities Commission presented a 15-year plan *Outline of Japan's space development programme*. This plan was revised at regular intervals and attempted to harmonize the different scientific, industrial, applications and engineering interests involved in the programme. This overall plan had some effect in producing an over-arching coordination between the two wings of the Japanese space programme.

At around the same time, the Space Activities Committee of industry proposed a 15-year plan involving 75 launchings and the spending of ¥1,479bn (€13.2bn) involving:

NASDA's launch centre at Tanegashima.

- a Japanese space laboratory as part of the American space shuttle programme, with equipment operated by Japanese astronauts;
- development of a fully hydrogen-powered launcher;
- a deep space programme with lunar sample return and probes to Venus, Mars, Jupiter and Saturn;
- participation in a future American space station;
- study of interstellar probes.

Whatever the degree of coordination, NASDA has always obtained the lion's share of Japanese space spending, generally around 80%. Between 1955 and 1973, ISAS received only a quarter of all space spending. In the present day, the proportion is much less (less than 8%).

NASDA'S LAUNCH SITE

NASDA moved quickly to set up its own launch site and plan the development of powerful, liquid-fuel rockets. The principal objective was to develop a rocket which could deliver a payload to 24-hr orbit—the position, 36 000 km above the Earth where a satellite takes one day to orbit the Earth and thus appears to hover all the time, the ideal

Tanegashima

Tanegashima launch centre.

location for communication and weather satellites—comsats and metsats. This is also called a geosynchronous orbit.

Some liquid-fuelled rockets had been tested on Nijima island in Tokyo bay in 1964 and another site for sounding rockets had been developed in Takesaki, Tanegashima island, in southern Japan in the course of 1966–68. Soon afters its establishment, NASDA developed the Osaki site just north of Takesaki on Tanegashima island as a base for its first rocket. Like their colleagues in Kagoshima, they were to suffer the same restrictions by the fishermen.

Tanegashima is 100 km south of Kagoshima and over 1000 km from Tokyo. It is a picturesque site, set in rolling hills alongside a rocky-sandy seashore where gentle waves roll in from the Pacific. The initial construction cost was ¥1.466bn (€13m). Tracking stations were set up at Katsuura, on the Boso peninsular near Tokyo and Okinawa.

NASDA'S ROCKET

When NASDA was first formed, the original intention had been for Japan to develop an indigenous powerful rocket capable of putting technology satellites into low Earth orbit and up to 100 kg into geostationary orbit. This was the Q rocket and was intended to make its first flight in 1974–75, using a liquid-propelled bottom stage in addition to a solid-fuel upper stage. However, NASDA changed its mind and decided instead to re-build the new Japanese launcher around licensed American technology, an approach which it reckoned would be much faster and cheaper. The Q rocket was dropped and the N rocket (N for Nippon) was approved in October 1970, based on the American *Thor* booster. The Mitsubishi company was charged with the N rocket, the engine and the first and second stages (on licence from Rockwell) and the third solid-fuel stage (on licence from Thiokol). The first stage motor was the American Mb-3. The second stage motor, the LE-3, was Japanese-built, with American assistance. Design and construction took place in the early 1970s.

FIRST LAUNCH OF N-1

The N-I was used to launch the first engineering test satellite (ETS) or *Kiku*, which means 'chrysanthemum' in Japanese. The concept of the ETS series was that *Kiku* satellites

Countries to reach geostationary orbit

Date	Country	Satellite
1965	United States	*Early Bird*
1974	Soviet Union	*Molniya IS*
1977	Japan	*Kiku-2*
1979	Europe	*Ariane*
1984	China	*Shiyang Weixing*

would test out new technologies that would later be applied to the Japanese space industry, mainly in the field of communications. The small 83-kg ETS-1, on 9 September 1975, was a test of the N-1 launcher and its tracking systems. The satellite was 80 cm in diameter and had 26 sides and entered an orbit swinging out to 1100 km.

Eighteen months later, the N-I achieved its prime objective when drum-shaped ETS-2 was the first Japanese satellite to reach geosynchronous orbit. With *Kiku-2*, Japan became only the third country in the world to reach geostationary orbit.

N-I (Mitsubishi)		
Length: 32.57 m		
Diameter: 2.44 m		
Weight: 90.26 tonnes		
Performance: 130 kg to geosynchronoous earth orbit (GEO)		
Stage	Engines	Thrust, kg
Strap-on	Two TX-354-5 solid	71 000
1	MB-3 liquid	77 000
2	LE-3 liquid	5 400
3	TE364-3 solid	6 800

Kiku-1

Ume-1 (*ume* means 'apricot') was put into orbit by the N-I on 29 February 1976 to monitor radio waves in the ionosphere and use the results to forecast short-wave radio communication conditions. Power failed after one month and the backup 141-kg model was put into 975–1224 km, 69°, orbit as *Ume-2* on 16 February 1978 where it worked successfully for five years.

FIRST ENGINEERING SATELLITES

The testing out of communications technologies was an early priority for the N-I rocket programme. ECS-1, or *Ayame-1*, was an experimental, Japanese-built comsat, launched on the N-I on 6 February 1979. There were reports later that it was lost when it collided with the third stage of the launcher.

ECS-2, *Ayame-2*, was launched on the N-I on 22 February 1980 and was a 260 kg, cylindrical, experimental comsat. Contact was lost when the apogee motor failed to fire, stranding it in a useless orbit.

Kiku-4 was a 385 kg, box-shaped experimental satellite to develop new solar arrays, attitude control systems and thermal protection. *Kiku-4* carried a vidicon camera for Earth images, a magnetic attitude control system and an ion engine.

N-I history	
9 Sep 1975	ETS-1, *Kiku-1*
29 Feb 1976	*Ume-1*
23 Feb 1977	ETS-2, *Kiku-2*
16 Feb 1978	*Ume-2*
6 Feb 1979	ECS-1, *Ayame-1*
22 Feb 1980	ECS-2, *Ayame-2*
3 Sep 1982	ETS-3, *Kiku-4*

Note: ETS-3 became Kiku-4; ETS-4 became Kiku-3

COMMUNICATIONS SATELLITES

An early priority in Japan was the development of communications services for telephone and television. In order to speed up the introduction of operational communication technology, NASDA encouraged Japanese companies to buy in expertise from abroad. The size of the satellite required to develop these services was bigger than what the N-I rocket launch, so the use of foreign rockets was permitted. Mitsubishi awarded Philco Ford a ¥3.15bn (€28m) contract for the launch of two experimental domestic communications satellites on American *Delta* rockets (*Sakura*); and Tokyo Shibaru Electric placed another ¥3.15bn (€28m) contract with General Electric for two experimental television satellites (*Yuri*), also to fly on American *Deltas*.

From then until 1989, communications satellites were developed through NASDA, the Telecommunications Advancement Organization of Japan and private companies operating under government regulation. Japanese communications satellites during this period may be classified into four main groups: the *Yuri* or BS series of direct broadcast satellites; the CS (Communication Satellite) or *Sakura* series of television broadcast satellites; telecommunications satellites of the JCSat series; and *Superbird*. The following were the main elements of the system.

Yuri series

BSE-1, which stands for Broadcasting Satellite Experimental, or *Yuri-1*, was launched on a *Delta* in April 1978 to test methods of transmitting colour TV to the Japanese islands and Okinawa. The tests were carried out by NASDA and the Japan Broadcasting Corporation, NHK, which had been researching direct broadcasting technology since 1965. BSE weighed 678 kg, was box-shaped and built by General Electric and Toshiba. With *Yuri*, Japan became the first country to experiment with direct television broadcast to the home. The overall cost of the experiment was ¥28.8bn (€257m).

Yuri communications satellite

The successors to *Yuri* were the BS-2A and BS-2B satellites whose aim was to develop operational direct broadcast colour television to small dish receivers and provide general and educational television to 400 000 households who did not have conventional access to television, mainly people living on the remoter islands. BS-2A and B were built by General Electric and Toshiba and operated by the Telecommunications Satellite Corporation (Telesat-Japan). Each was 670 kg, used Thiokol Star apogee kick motors, and was equipped with 8.9 m long solar arrays. Home receivers were generally less than

1 m diameter with a converter unit. Nothing was left to chance and in advance of the tests, the home receivers were subjected to the typhoons of the southern islands and the harsh winter snows of the northern island of Hokkaido. The BS-2 series was also used to test out high-definition television. BS-2A was switched on over Borneo in May 1983 for the first time and 420 000 viewers on the southern Japanese islands were able to receive direct satellite television for the first time. However, two of its three transponders failed within three months and full service had to await BS-2B, launched in February 1986.

Six years after the introduction of the BS-2, *Yuri-2* series, the BS-3, *Yuri-3* set was brought in. About 83% of its components were built in Japan. *Yuri-3A* was launched on H-I from Tanegashima in August 1990 but suffered a partial solar array failure. It was eventually taken out of station in April 1998, its motor being fired to put it into a graveyard orbit.

Sakura series

Sakura was an experimental medium-capacity geostationary telephone and television satellite, the first being launched by American *Delta* rocket on 15 December 1977. It was 3.51 m tall and 2.18 m diameter, weighing 676 kg and carrying six transponders. It was the first satellite to use quasi-millimetre waves to transmit signals.

This paved the way for an operational system providing general area coverage. The *Sakura-2* series followed in 1983: these were the first comsats to use the high-capacity K-band frequencies (20–30 GHz). *Sakura-2A* was launched 4 February 1983 on the H-I and *2B* on 5 August. Built by Ford Aerospace and Mitsubishi, each could handle 4000 telephone calls at a time. *Sakura-2A* and *2B* were 355 kg comsats carrying telephone, television and data channels, designed to link the Japanese islands. These were the first operational Japanese domestic communications satellite.

Sakura-3A, launched on H-I on 19 February 1988, was put into orbit to develop more advanced satellite communications technologies. *Sakura-3A* was 3 m long, 2.2 m in diameter cylinder, weight 1100 kg. Among its many functions, it provided datalinks for the construction ministry and the National Police Agency. *Sakura-3B* followed in quick succession, on 16 September that year.

JCSat series

JCSat are commercial satellites developed by the Tokyo-based Japanese Communications Satellite Company, a joint venture between Itoh, Mitsui and other large Japanese trading companies formed in April 1985 to bring telephone and television services to business and private users in Japan. JCSat 1 was launched by *Ariane 4* on 6 March 1989 into 24-hr orbit. JCSat-2, launched by American *Titan 3* in 1990, involved the American satellite builder Hughes, and the satellite itself was based on the proven Hughes 393 model—a 10 m tall, 3.66-m diameter drum with a 2.4-m antenna able to produce a beam to cover all the Japanese islands. JCSat 3 was a Hughes 601, launched by *Atlas 2AS* in 1995, JCSat 4 using the same launcher in 1997.

Superbird series

Superbird A, set up to rival JCSat, was launched into 24-h orbit on *Ariane 4* on 5 June 1989. However, *Superbird* failed in December 1989 when its attitude control system malfunctioned, the oxidizer apparently leaking overboard. Superbird is owned by the Space Communications Corporation, a company developed by Mitsubishi. *Superbird B* was launched by *Ariane* on 26 February 1992 and *Superbird C* by an American to rocket on 27 July 1997.

INTRODUCING THE N-II

The N-I had a number of drawbacks. First, it was able to send only about 130 kg to geostationary orbit. This was fine for experimental purposes, but most operational communications satellites by the late 1970s required a lifting capacity of at least twice that amount. Second, by the time the N-I was developed by the Japanese it was already well out of date. The N-I first flew in 1975, but most of its technology (e.g. the guidance system) dated to the 1960s. No sooner was it flying than Japanese space experts were thinking of the need for a more powerful vehicle and this was approved in September 1976. Again, a licensing arrangement was entered into with the United States. Prime contractor was Mitsubishi Heavy Industries.

These licensing arrangements have rarely been discussed much in public, but they appear to be a running sore to at least some Japanese. Whilst the licences forbade the transfer of the technology to third parties or countries, a normal and reasonable condition, the operation of the agreements also effectively prohibited Japan from offering its American-derived launchers on the world commercial market. Some of the technology associated with the licences was classified, and the Japanese were not allowed to know about some of the components which they were themselves operating. Whenever problems arose in the N-I rocket, American technicians had to be called on to fix faults [2].

1981 saw the introduction of the N-II launcher. For the Japanese, the N-II was an essential part of their efforts to place larger satellites into 24-hr orbit for the 1980s and reduce reliance on American *Deltas*.

The N-II more than doubled Japan's ability to reach 24-hr orbit, from 130 kg to 350 kg. Just as the N-I had been based on an earlier American launcher, the *Thor*, the N-II was based on a more recent version of the same *Thor*, the *Thor-Delta*. Compared to the N-I, the N-II featured a number of improvements, principally nine solid-fuel strap-ons (rather than three), a longer first stage with 34% more capacity, and a new motor for the second stage (the Aerojet AJ10-118F). The strap-ons burned for 38 sec and were dumped 85 sec into the mission. The third stage consisted of a solid fuel Thiokol TE-M-364-4 motor, with a propellant weight of 1.1 tonnes and thrust for 44 sec. The third stage was supplied directly from the United States rather than built in Japan. The N-II was built in Mitsubishi's factory in Nagoya.

The first launch of the N-II was *Kiku-3*, a 640-kg cylindrical satellite launched on 11 February 1981, carrying a pulse plasma engine. Just before *Kiku* went up, five members of NASDA's ruling council went to a Tokyo Shinto shrine to pray. The mission objectives were to test the operation of the test satellite, the generation of electrical power and the operation of the N-II launcher. Later, the rocket was used for launching weather and communications satellites to 24-hr orbit, concluding with Japan's first marine observation satellite.

N-II (Mitsubishi)
Length: 35.35 m
Diameter: 2.44 m
Weight: 135 tonnes
Capability: 350 kg to GEO

Stage	Engines	Thrust, kg
Strap-on	Nine TX-354-5 solid	213 300
1	MB-3 liquid	77 000
2	AJ-10-118F liquid	4 400
3	TE364-3 solid	6 800

WATCHING EARTH'S WEATHER

Paralleling the development of communications satellites, an early achievement of NASDA's space programme was the orbiting of a system of weather satellites. These were especially important for Japan, being both an island nation and one periodically affected by typhoons.

Japan's first Geostationary Meteorological Satellite (GMS), built by Nippon Electric and Hughes for NASDA, was launched from Cape Canaveral on a *Delta 2914* on 14 July 1977 and was positioned at 140°E, just south of Tokyo itself. Called *Himawari*, or 'sunflower', the 130-kg cylindrical satellite was 3 m long, 1.9 m in diameter and covered with solar cells at the side. *Himawari* was Japan's contribution to the Global Atmospheric Research Programme, or GARP, sponsored by the World Meteorological Organization (Europe contributed *Meteosat* and the United States two GOES satellites). The satellite itself, and its successors, was based on the American GOES weather satellite model. *Himawari's* aim was to carry out a global weather watch, collect and distribute weather data and monitor solar particles. *Himawari* was stationed in such a way that it could scan the planet from Hawaii in the east to Pakistan in the west with visible and infrared scanners. Its main instrument was a Visible and Infrared Spin-scan Radiometer, providing high resolution visible images of the Earth every 30 min. It also carried a Space Environment Monitor to measure ionized gases entering the Earth's atmosphere from the sun.

A second generation GMS was launched four years later, this time by the domestic N-II launcher. GM-2, *Himawari-2*, was 20 kg lighter, owing to the slightly lower capacity of the N-II, but it made up for the reduction by the use of lighter materials. GMS-2, like its predecessor, was drum-shaped, with observation camera and vizor on top, on which rested the two spacecraft antenna. A Space Environment Monitor recorded solar particles and their effect on Earth's communications. The main instrument was the Visible Infrared Spin-scan Radiometer, which scanned the entire planet: its first picture, relayed in early September 1981, showed the swirling clouds of the Pacific ocean, Australia showing clearly through, filling the bottom left side of the globe. Placed at

N-II history	
11 Feb 1981	ETS-4/*Kiku-3*
11 Aug 1981	GMS-2/*Himawari-2*
4 Feb 1983	CS-2A/*Sakura-2A*
5 Aug 1983	CS-2B/*Sakura-2B*
23 Jan 1984	BS-2A/*Yuri-2A*
3 Aug 1984	GMS-3/*Himawari-3*
12 Feb 1986	BS-2B/*Yuri-2B*
19 Feb 1987	MOS-1/*Momo*

140°E, GMS-2 transmitted pictures eight times a day until November 1983 when its electrical motor began to break down. When GMS-2 broke down, the Japanese weather agency took GMS-1 out of retirement at 160°E to take over its work until GMS-3 could be launched. GMS-3, *Himarwari-3*, built by Hughes for Nippon Electric NEC, followed in August 1984, providing, every 30 min, infrared and visible light views of Japan, China, south-east Asia and Australia.

The first attempt to launch GMS-4, *Himawari-4* from Tanegashima on H-I failed on 8 August 1989 when there was a rare pad abort. The computer detected a potential valve failure in the first stage motor just after it ignited and halted the countdown just in time. A second attempt a month later on 5 September succeeded. The satellite was placed in geostationary transfer orbit 16 hr after launch and eventually manoeuvred into position at 150°E.

The GMS series concluded with GM-5, *Himawari-5* in 1995, which replaced GMS-4. It carried a visible infrared spin-scan radiometer providing full pictures of the Earth's disc every 25 min in one visible and three infrared bands and was designed to measure sea temperatures and water vapour more precisely than ever before. The instrument had a resolution of 1.25 km (visible light) and 5 km (infrared light) respectively.

H-ROCKET: INTRODUCING LIQUID HYDROGEN

The next series of rockets, the H series, attempted to build on the success of the American-licensed N-I and N-II. The purpose of the H series was to double again the payload to 24-hr orbit, this time to 550 kg, to replace more American technology with equipment designed and built in Japan, to make Japan independent as a launching country and to offer launches at internationally competitive rates. This time, 80% of the H-I was made in Japan. Approval to begin the project was given in February 1981.

Central to the H rocket concept was the introduction of a hydrogen-powered cryogenic upper stage. Hydrogen-powered upper stages give considerable extra boost for payloads destined for 24-hr orbit or deep space. The technology is called cryogenic because it involves handling the hydrogen fuel at extremely low temperatures (hydrogen boils at −183°C). The low temperatures, combined with the explosiveness of hydrogen, posed difficult engineering challenges. The United States had developed the first hydrogen-

Mu launch centre at Kagoshima

GMS series		
GMS-1/*Himawari-1*	14 Jul 1977	*Delta*
GMS-2/*Himawari-2*	11 Aug 1981	N-II
GMS-3/*Himawari-3*	3 Aug 1984	N-II
GMS-4/*Himawari-4*	5 Sep 1989	H-I
GMS-5/*Himawari-5*	18 Mar 1995	H-II

power upper stage, the *Centaur*, in the 1960s, though not without difficulty nor some spectacular explosions.

For this dangerous enterprise, the two wings of the Japanese space programme, ISAS and NASDA, came together. As far back as 1972, Professor M. Nagatomo of ISAS had proposed a seven-tonne cryogenic upper stage for the *Mu* solid-fuel rocket; NASDA, for its part, wished to reduce its dependence on American-licensed rockets and motors. ISAS had carried out model tests of its engine in the late 1970s at the Noshiro test centre when the Space Activities Commission asked ISAS to build a larger engine in cooperation with NASDA. The go-ahead for a cryogenic, hydrogen power upper stage was given in 1982. The new, restartable motor was called the LE-5, the contract being awarded to

Mitsubishi. The 255-kg LE-5 was designed to develop 10.5 tonnes of thrust, with a specific impulse of 447 sec and a combustion pressure of 37 atmospheres. This became the basis of the second stage of the H-I.

	H-I (Rocket Systems Corp.)	
	Length: 40.3 m Diameter: 2.44 m Weight: 139.9 tonnes Capability: 550 kg to GEO	
Stage	Engines	Thrust, kg
Strap-on	Nine Castor 2 solid	150 000
1	MB-3 liquid	88 000
2	LE-5 liquid hydrogen	10 500
3	Solid	7 900

In a typical launch profile, six of the H-I's solid rocket strap-ons ignited at lift-off. These burned out at an altitude of 4 km when the other three strap-ons ignited. All nine were jettisoned 85 sec into the mission to tumble into the Pacific Ocean. The first stage burned out at 4 min 30 sec. Eight seconds later, now soaring above the atmosphere, the second stage would light up. The payload fairing at the top would come off 5 min 14 sec into the mission. The liquid hydrogen second stage would typically burn for 363 sec bringing the vehicle to a velocity of 7.8 km/sec. After a 22-sec coast, the solid third stage would light for a burn of about a minute to achieve a geostationary transfer orbit, following which the payload would separate 26 min after take-off. A further minute's burn would circularize the orbit.

The H-I made its first flight on 13 August 1986. A splendid white rocket, with the Japanese flag and 'Nippon' written in black on its side, it took into orbit three payloads on its first mission—an experimental geodetic satellite, an amateur radio satellite and a magnetic bearing flywheel experimental system. The experimental geodetic payload, *Ajisei* (meaning 'hydrangea') was a 685 kg passive satellite designed to improve triangulation measurements of the surface of the Earth from an altitude of 1500 km. Built by Kawasaki, it entered a circular orbit of 1479 km. In the shape of a 2.15 m ball, it was covered with 120 glassy reflectors designed to beam back both laser and light beams. Launched with *Ajisei* was *Fuji*, a 26-sided, 50-kg amateur radio satellite only 0.5 m across.

Kiku-5, a 550-kg satellite launched by H-I in August 1987, was a technology development satellite to test the use of C-band and L-band transponders for ships and aircraft and was placed in geosynchronous orbit at 150°E. It was the first to use an indigenous kick motor. In an experiment in Pacific regional cooperation, it was used to transmit educational programmes to Fiji and Papua New Guinea. The apogee motor developed by

ETS-5 was later used for the BS-3 series of broadcasting satellites, a good example of how the engineering programme paved the way for a commercial application.

H-I history

13 Aug 1986	EGS/*Ajisei*, with *Fuji-1*
27 Aug 1987	ETS-5/*Kiku-5*
19 Feb 1988	CS-3A/*Sakura-3A*
16 Sep 1988	CS-3B/*Sakura-3B*
5 Sep 1989	GMS-4/*Himawari-4*
7 Feb 1990	MOS-1B/*Momo-1B*
25 Aug 1990	BS-3A/*Yuri-3A*
25 Aug 1991	BS-3B/*Yuri-3B*
11 Feb 1992	JERS-1/*Fuyo*

SOUNDING ROCKETS

Although sounding rockets paved the way for Japan's first orbital missions, they did not become redundant when orbital spaceflight became commonplace. Between the start of the space age and 1996, Japan flew 325 sounding rockets. The *Kappa-9M* series from the 1960s continued in use for many years. Ninety were fired altogether. A sounding rocket campaign was conducted with the United States at Wallops Island in 1967.

ISAS continued to use sounding rockets for microgravity experiments. The main sounding rocket developed was the Nissan-built S-520 which first flew in 1980 and replaced the *Kappa-M*. It used the ISAS Kagoshima site but has also been fired from the Andoya range in Norway. The rocket weighs 2.285 tonnes, is nearly 9 m long and can reach an altitude of 350 km. It is launched from a platform through the roof of an enclosed pad, reminiscent of the way a telescope sticks through the dome of an observatory. ISAS also uses: the MT-135, the smallest sounding rocket, developed by Professor F. Tamaki, used to sample the middle atmosphere for ozone depletion; the S-310, a medium-size sounding rocket; and the SS-520, which has two stages and can send an experimental payload up as far as 1000 km.

The S-310 and the S-520 have been launched abroad—in Antarctica (the Japanese base at Showa) and Andoya, Norway; and launches of the SS-520 have been scheduled for Spitsbergen. On 18 February 1983, a S-520 sounding rocket was used for a Microwave Energy Transmission in Space (METS) experiment, to test the possibility of transmitting solar energy to power the electricity grid on Earth.

For the weather service, the MT-135 rocket, made by Nissan, is launched almost weekly from the Japanese Meteorological Agency's weather station at Ryori on the north-east coast. The Department of Education in Japan was required to make annual surveys of the problem of ozone in the atmosphere.

Japanese sounding rockets				
	MT-135	S-310	S-520	SS-520
Length	3.4 m	7.1 m	8.6 m	9.65 m
Diameter	13.5 cm	31 cm	52 cm	52 cm
Weight	70 kg	700 kg	2.2 tonnes	2.6 tonnes
Max. payload	6 kg	50 kg	150 kg	140 kg
Altitude	60 km	210 km	270 km	1000 km

CONCLUSIONS

By the this time, the Japanese space programme had made considerable gains. The progress achieved by the ISAS scientific satellites had been consolidated. Systems of weather and communications satellites were being established. Three licensed rockets had been developed and flown, several carrying early engineering satellites. Even as NASDA made rapid strides, ISAS had progressed well beyond its original programme of small, Earth-orbiting scientific satellites. Not only were more ambitious, specialized missions undertaken, but ISAS had set its sights on the Moon and planets.

Three Japanese sounding rockets—S-520, S-310 and S-310 in Antarctica

3

Scientific programmes

Throughout the 1970s, 1980s and 1990s, ISAS continued its development of Japanese scientific space programmes. Although in terms of budgets and rocket lifting power it was the poorer cousin of NASDA, some of its missions were spectacular and attracted more public interest.

On 14 April 1981, ISAS changed its name from the Institute of Space and *Aero*nautical Sciences (ISAS) to the Institute of Space and *Astro*nautical Sciences (conveniently also ISAS) and became a national, inter-university research institute, independent of the University of Tokyo and answerable to the Ministry of Education, Science & Culture. It had become too big for just one university and soon set up its own dedicated campus at Sagamihara. The first director of the reformed institute was Professor D. Mori.

THIRD *MU* VERSION—THE NEW *MU-3H*

The programme of regular satellite launchings continued with the small, solid-fuel *Mu* launcher. These satellite programmes now pursued distinct themes: for example, the Astro series made astronomical observations, the Solar series studied the sun, Exos the atmosphere and so on. As a result, some had a series designator in addition to their Japanese name. Efforts continued to improve on the performance of the *Mu* rocket. For the *Mu-3H*, the first stage was lengthened.

Kyokko (meaning 'aurora') was a 103-kg satellite flown in 1978 to investigate aurora, while *Jikiken* (meaning 'magnetosphere') was a 100-kg research satellite carrying equipment to study plasma, charged particles, electric and magnetic fields, launched in September 1978.

FOURTH *MU* VERSION—THE *MU-3S*

Following *Jikiken*, a new version of the *Mu-3* was introduced, the *Mu-3S*, with first-stage vector control, improved accuracy of orbital insertion and a 50% increase in payload (300 kg).

Mu-3H

Length: 23.8 m
Diameter: 1.41 m
Weight: 48.7 tonnes
Performance: 200 kg to Earth orbit

Stage	Engines	Thrust, kg
Strap-on	Eight SB-310	77 600
1	M-13	97 300
2	M-22-TVC	28 400
3	M-3A	5 800

M-3H history
19 Feb 1977   *Tansei 3*
 4 Feb 1978   Exos-A/*Kyokko*
16 Sep 1978   Exos-B/*Jikiken*

Mu-3S

Length: 23.8 m
Diameter: 1.41 m
Weight: 48.9 tonnes
Performance: 300 kg to Earth orbit

Stage	Engines	Thrust, kg
Strap-on	Eight SB-310	77 600
1	M-13-TVC	104 000
2	M-22-TVC	28 400
3	M-3A	5 900

The *Mu-3S* was used to make almost annual satellite launchings in the early 1980s. *Hinotori* was a small scientific satellite launched 21 February 1981 with an X-ray telescope and spectrograph for solar studies. It produced significant results in X-ray and neutron star astrophysics.

The purpose of Astro-B, or *Tenma* ('flying horse') was to image celestial X-ray sources, such as nebulae, galaxies and bursts. *Tenma* consisted of a 89.5 cm by 110.4 cm box weighing 218 kg, with four solar panels fitted at the base providing 150 W of electrical power. *Tenma* transmitted to the ground both in real time and through stored data five orbits a day out of its 15 daily passes. *Tenma* was Japan's eighth scientific satellite and second X-ray satellite, carrying five instrument to observe X-rays from stars and galaxies. Launched on 20 February 1983, the instruments were ten scintillation proportional counters, transient X-ray source monitor, X-ray focusing collector, radiation belt monitor and gamma burst detector.

Mu-3S	
17 Feb 1980	*Tansei 4*
21 Feb 1981	*Hinotori*
20 Feb 1983	Astro-B/*Tenma*
14 Feb 1984	Exos-C/*Ohzora*

Exos-C or *Ohzora* ('big sky') was part of a middle-atmosphere programme. From an altitude of 300 to 800 km, it used optical instruments to observe phenomena in the Earth's atmosphere from 10 to 130 km. One particular area of study was the South Atlantic Geomagnetic Anomaly and an alarm system was fitted to note whenever the satellite overflew the anomaly. The satellite weighed 180 kg and was a cuboid, 1 m each side, with four solar panels.

PROBES TO COMET HALLEY

1985–86 marked the first year since the start of the space age for a major regular comet to make an approach to the inner solar system. Halley, the most famous of all the comets, named after astronomer royal Edmond Halley who first characterized cometary orbits, had in fact first been observed as far back as 240 BC. Since then it had been observed 28 times, every 76 years (no record of its appearance in 164 BC has come down to us). Several countries prepared space missions to intercept or pass close to Comet Halley in the period around its closest approach to the Sun, or perihelion passage, due on 9 February 1986.

The United States, by this time the leader in interplanetary exploration, was, ironically the only country not to send a probe to Comet Halley, though a small *Explorer*-class scientific satellite was sent into its distant tail. American spending on deep-space missions was very low in the late 1970s and a mission to Comet Halley was one of many casualties of the glacial domestic financial atmosphere of the period. The Soviet Union mounted a spectacular double mission, sending probes *Vega-1* and 2 to Venus (where they dropped probes and balloons) before altering course to intercept the comet. Most dramatic of all, the European Space Agency fired the *Giotto* probe right into the head of the comet.

For Japan, this was the opportunity to organize its first deep-space mission. Two missions were organized—a pathfinder and a main probe. The pathfinder, MS-T5, flew eight months in advance of the main probe, Planet A, both spacecraft being built by the Nippon Electric Co. of Yokohama. Planet A was to close to within 10 000 km of the comet. These would be the first deep-space probes to use solid-fuelled rockets. Because

Probes to Comet Halley—the *Mu-3SII*

Comet Halley had not arranged its visit to the inner solar system to suit the southern Japanese fishermen, delicate negotiations took place for a breach in the normal regulations so as to permit out-of-season launches to chase the comet.

Planet A was a small probe, able to carry only 10 kg of instrumentation. The spacecraft was drum-shaped, 70 cm in height and 1.4 m in diameter, weighting 135 kg, with a solar array around the side, charge-coupled device cameras, scientific instruments (principally charged particle collector) and an 80-cm diameter high-gain antenna on top. The camera was designed to take pictures to a resolution of 30 km during the encounter. Planet A had low-thrust gas jets which gave it the ability to carry out limited course corrections—10 kg of hydrazine propellant for six 3 N thrusters for trajectory correction and attitude and to settle the spin of the spacecraft. Two thousand solar cells provided from between 67 and 104 W of electrical power, depending on the distance from the Sun.

MS-T5 was almost identical, though weighing in slightly heavier at 141 kg, also with 2000 solar cells for electrical power. MS-T5 carried three experiments to detect plasma wave instability, measure the solar wind and analyse the structures of the interplanetary

Usuda Deep Space Tracking Centre

magnetic field. Mission director was Professor K. Hirao of the Planetary Research Division of the ISAS Tokyo laboratory. The spacecraft were built by Mitsubishi, using lightweight carbon-fibre-reinforced plastic and assembled at the ISAS Sagamihara space facility. The probes were the first to use a new launch building at the Kagoshima space centre.

A special version of the *Mu-3* launcher was devised, the *Mu-3SII*, with two strap-on boosters to give additional thrust. The *Mu-3SII* weighed 61 tonnes, 12 tonnes more than the standard model. The *SII* model was a substantial improvement on the S version, being heavier (61 tonnes compared to 48.7 tonnes), longer (27.8 m compared to 23.8 m), with a doubled payload (up from 300 to 770 kg). The *SII* benefited from the first stage of the *S* but had improved second and third stages with movable nozzles and an optional kick fourth stage. Although introduced for one mission, it became the basis of a series of later scientific flights.

The probes were sent directly into solar orbit, without the use of an Earth parking orbit. MS-T5 was launched from Kagoshima on 8 January 1985, followed by Planet A on

Mu-3SII (Nissan)		
Length: 27.8 m		
Weight: 61 tonnes		
Diameter: 1.41 m		
Performance: 770 kg to Earth orbit		
Stage	Engines	Thrust, kg
Strap-on	Two SB-375 solid	30 424
1	M-13 solid	114 285
2	M-23 solid	52 960
3	M-38 solid	12 040

19 August 1985 at the start of a 20-day window. Three days into its mission, MS-T5 made the first of a series of mid-course corrections with its small pulse jets. MS-T5, renamed *Sakigake* ('pioneer'), was able to concentrate on the area downstream of the Sun and study the solar wind, magnetic fields and plasma waves, approaching to within 7 million kilometres of the comet on 12 March 1986.

Planet A, renamed *Suisei*, made its closest approach, 200 000 km from the head of the comet, on 8 March 1986, just a month after the perihelion passage. *Suisei's* instruments noted clearly the moment when the little spacecraft crossed the bow shock generated by the comet against the solar wind. The camera was able to focus on the comet's nucleus and its hydrogen cloud for a month following the interception and found the rotation period for the comet. The two probes were tracked by a new 64 m deep-space antenna constructed at Usuda, in the radio-quiet Nagano valley 170 km north-west of Tokyo, completed on 31 October 1984. After the mission was over, Usuda tracked the American *Voyager 2* flyby of the planet Neptune in 1989.

Seven years later *Sakigake's* orbit intercepted that of Earth, the small spacecraft passing 90 000 km over the Indian Ocean. Its motor was used again to reset the spacecraft's orbit in such a way that it could better study the Earth's magnetic field and the solar wind.

The Japanese probes were probably the least publicized of the earthly armada which flew to Halley in 1986. They did attract considerable public interest in Japan itself and the Tokyo Broadcasting Service made 16 television programmes on the probes.

The *Mu-3SII* went on to become the most used rocket of the *Mu-3* series.

In addition to the *Mu-3SII*-launched satellites of this period, the Japanese–American *Geotail* was launched by American *Delta* on 24 July 1992, to study the structure and dynamics of the Earth's magnetic tail. *Geotail* used lunar swingbys to steer itself repeatedly into the tail of Earth's magnetosphere. *Geotail* was a drum-shaped spacecraft with two 6-m masts and two 100-m antennae, designed to measure the magnetic field, electric field, plasma, energetic particles and plasma waves.

Mu-3II history

8 Jan 1985	MS-T5/*Sakigake*
19 Aug 1985	Planet A/*Suisei*
5 Feb 1987	Astro-C/*Ginga*
21 Feb 1989	Exos-D/*Akebono*
24 Jan 1990	Muses A—*Hiten/Hagoromo*
30 Aug 1991	Solar-A/*Yohkoh*
20 Feb 1993	Astro-D/*Asuka*
15 Feb 1995	*Express*

MU-3SII SCIENTIFIC MISSIONS

Exos-D, renamed *Akebouo* ('dawn'), but sometimes also known as ASCA (Advanced Satellite for Cosmology & Astrophysics) entered a highly elliptical orbit out to 10 460 km at 75° on 21 February 1989 and was designed to study particle acceleration in the southern lights, the *aurora australis*. Canada participated in the programme.

The 430-kg Astro-C was developed in collaboration with British scientists at the University of Leicester and the Rutherford Appleton Laboratory. Astro-C, known as *Ginga*, was launched in February 1987 and carried what was then the largest satellite-borne X-ray detector weighing over 100 kg. Its aim was, over five years, to obtain important new information on X-rays released from the vicinity of neutron stars and black holes.

Astro-D, known as *Asuka*, was a 420-kg X-ray satellite launched on 20 February 1993 carrying four telescopes, the successor to *Ginga* which had completed its mission at the end of 1991. *Asuka* entered orbit of 538–647 km, 96 min. It was a 7.7 m tall, 1.3-m diameter box with 2.8 m panels to provide electrical power for five to six years. The four X-ray telescopes were ten times more powerful than those carried on *Ginga* and could pick up light from objects 10 billion light years' distant, rays which were normally absorbed by the Earth's atmosphere and could not be observed on the ground. Its X-ray camera was designed to provide clearer, sharper images than any previous X-ray camera. *Asuka* carried an American telescope called BBXRT, Broad Band X-ray Telescope, which had flown in space earlier on the American Astro-1 spacelab mission and had provided excellent results on active galaxies, hot interstellar gases, supernovae, black holes and quasars. Its transmitted images of the structure of the bright supernova SB1006 and of dark matter lying between galaxies. *Asuka* was the fourth of the programme of X-ray satellites dating back to *Hakucho*. Between 1993 and 1995, *Asuka* found 119 distant active galaxies and faint X-ray sources, compiling a map of the north galactic pole.

Astro-E launched on the *Mu-5* rocket, with two instruments for soft X-ray observations and a large counter array for hard X-ray observations, designed between them to ensure both a wide range and variety of sources and high sensitivity. 55 sec after liftoff on 10 February 2000, The *Mu-5* veered off course when its rocket nozzle failed. Astro-E burned up.

Japan's X-ray satellites
Astro-A   *Hakucho*
Astro-B   *Tenma*
Astro-C   *Ginga*
Astro-D   *Asuka*

Planned
2003   Astro-F
to be given a Japanese name on entering orbit.

Astro-F will be an infrared astronomy satellite with American cooperation to survey large areas of the sky, providing large data sets from which scientists hope to learn about the formation and evolution of star, galaxies, brown dwarfs and distant planetary systems. Astro-F, also called the Infrared Imaging Surveyor, will weigh 870 kg and take a 750 km orbit around the Earth. It will use mechanical cooling techniques designed to keep the telescope functioning for much longer than has been the case with previous infrared telescopes. The telescope will be cooled at the extraordinary temperature of –267°C, just 6°C above absolute zero. Astro-F is cylindrical with T-shaped solar panels at the top and will make the most detailed X-ray survey of the sky in 25 years, one designed to shed fresh light on the evolution of galaxies, the formation of stars and planets, brown dwarfs and dark matter.

SOLAR STUDIES

Japan launched the Solar A *Yohkoh* ('sunlight') probe on 30 August 1991, designed to study solar activity on a three-year assignment, entering an orbit of 519–787 km, 31°. The 390-kg satellite had four instruments to study the high-energy phenomena of solar flares: two telescopes (NASA built the soft X-ray telescope and the Japanese the hard X-ray telescope) and two spectrometers (the British Bragg Crystal spectrometer and the Japanese wide-band spectrometer). The Bragg instrument was built by the Mullard Space Science Laboratory and Rutherford Appleton Laboratory. *Yohkoh* was targeted to make observations during the solar maximum but proved equally valuable during the solar minimum of 1996 and the subsequent resurgence of activity. Pictures of the Sun during the maximum period showed a bright yellow, red and black surface gushing hydrogen; by contrast, there was only a glowing rim a few years later.

The *Yohkoh* mission went so well that in 1998 Britain and Japan signed up for a successor mission due in 2004 (Solar B). Solar B will be a 870-kg solar observatory which will be put in a 600-km circular orbit. Using X-ray and visible band instruments and a high-resolution solar telescope, it will study the relationship between the magnetic field and high-temperature plasma on the surface of the Sun.

Japan's solar missions	
1981	*Hinotori*
1991	Solar A/*Yohkoh*
2004	Solar B (planned)

JAPAN'S FIRST MOONSHOT: MUSES A

In 1982, ISAS began plans for a lunar orbiter of 650 kg to make a physical and chemical survey of the Moon. The success of *Sagikake* and *Suisei* four years later gave Japan the confidence to consider further missions outside Earth orbit. In 1987, ISAS formally sought approval from the Space Activities Commission for its first Moon probe to start a wide-ranging programme for lunar, planetary and cometary exploration. The project was called Muses A and the project manager was Professor Kuninori Uesugi. It was ready within three years.

Japan' first Moon probe was launch from Kagoshima on 24 January 1990. It was the first spacecraft launched to the Moon by any country since the Soviet Union's *Luna 24* had landed in the Sea of Crisis in August 1976 to bring rock samples back to Earth. Muses A made Japan the third country to launch a Moon rocket. The launch was the fifth of the *Mu-3-SII*. A new, 20-m antenna was built at Kagoshima to track Muses A.

The ¥4.3bn (€38m) Muses A project comprised two spacecraft:

— a 193-kg mother craft, 1.4 m in diameter, 79 cm high, called *Hiten*. Its weight included 42 kg of hydrazine fuel using 12 thrusters. Solar cells provided 100 W of power.
— a multisided 11-kg lunar orbiter, 40 cm in diameter, 27 cm high, called *Hagoromo*.

The mother craft was called *Hiten* after one of the subjects of Buddha, whose duty was to play music in heaven. It was a drum-shaped object with two antennae underneath and the little lunar orbiter placed on top. The only scientific instrument carried was a micro-meteoroid detector made in Munich, Germany. The mission was more an engineering than scientific one.

The first attempt to launch Muses A on 23 January reached 18 seconds before launch when technical problems halted the countdown. However, it got away the following day, and the *Mu-3SII* put the spacecraft into 240–400 km parking orbit around the Earth. After two circuits the solid rocket motor fired to place the spacecraft in a highly elliptical Earth orbit looping out to 16 000 km from the Moon two months later. In fact, the impulse given to *Hiten* was 50 sec less than planned, which meant that *Hiten* had to take an extra orbit to reach the Moon. Worse, during the translunar coast the transmitter on the *Hagoromo* subsatellite failed, which meant that no signals would be returned from lunar orbit.

On 18 March, 14 900 km behind the Moon and 54 days after launch, the mother craft released the orbiter which fired its own motor so as to enter lunar orbit. *Hagoromo* entered lunar orbit of 7400 by 20 000 km. The Kiso tracking observatory at the University of Tokyo was able, with its fine Schmidt camera, to spot the engine burning and confirm that *Hagoromo* had indeed entered lunar orbit.

Hiten meantime continued in its slow, lazy, Earth–Moon curving orbit. By October 1990, the distance from Earth stretched to 1.34 million kilometres. On 19 March 1991 its return trajectory took it 120 km back into the Earth's upper atmosphere to perform what was in effect the first-ever aerobraking manoeuvre at high velocity. Japanese mission controllers decided, in consultation with the American Jet Propulsion Laboratory in California, to take advantage of the fact that *Hiten* still had residual fuel so as to devise an extended mission. The aerobraking had reduced the apogee by 8600 km and a week later by a further 14 000 km. This swung *Hiten's* orbit back to the Moon in preparation, several orbits later, for an eventual lunar capture on 15 February 1992. *Hiten* entered a lunar orbit of 422 by 49 200 km, one amended by a plane change three months later. The orbit was not a stable one and the small spacecraft began to spiral inward. Rather than let *Hiten* crash at will, ground controllers used the very last fuel to guide *Hiten* to its final impact, so *Hiten* eventually hit the Moon near the crater Furnelius on 10 April the following year at 38°S latitude, 5°E longitude. Japan had become (1993) the third nation to hit the Moon after the Soviet Union (1959) and the United States (1962); and the third to enter lunar orbit after the Soviet Union and the United States (both 1966).

EXPRESS—FROM PACIFIC SEACOAST TO THE JUNGLES OF AFRICA

Last of the *Mu-3SII* launches was *Express*. The story of the *Express* capsule is one of the strangest in the Japanese space programme, one involving five other countries: Germany, Russia, Australia, Britain and Ghana. The German company OKB, in conjunction with the German space agency DARA had the idea of flying some microgravity experiments and returning them to Earth. The completion of the design studies coincided with a period of retrenchment in German public spending, not least in the space industry. The price tag became unaffordable.

In an effort to cut and spread costs, DARA sought out international partners. It persuaded the Khrunichev company in Russia to let them have a reentry capsule used for its now-defunct military Fractional Orbital Bombardment System (FOBS). This was a project, tested in the 1960s, to send small warheads around the earth on a three-quarter orbit to descend on the United States from Mexico.

The *Express* capsule, weighing 405 kg, had sufficient volume to carry up to 165 kg of payload. It was fitted with six experiments, three from Japan and three from Germany. These were CATEX (a Japanese experiment to produce crystals); HIPMEX and RTEX (Japanese reentry materials); CETEX, PYREX and REFLEX (German tests of ceramic tiles, train-braking systems and high-temperature incinerators). Accordingly, DARA fitted out the cabin's interior with its experimental packages and data-handling system. A connection was set up to link the German and Russian computers to ensure effective control over the spacecraft. Japan offered a free launch in exchange for full access to the research results.

The weight of the cabin was at the extreme limit of the performance of the *Mu-3SII* launcher. A 365-kg service module, connected to the research cabin, containing an attitude control system, retrorocket system, transmitters and battery supplies for up to a week, brought the weight to 762 kg, with an overall size of 2.3 m in length and 1.4 m in diameter.

Where to land the cabin posed a further challenge. Fortunately, the Australians came to the rescue, for the Australian Space Office had long been searching for ways to bring the Woomera Test Facility back into business. It was therefore agreed to recover the satellite near the Woomera launching base in the Australian desert some 5.5 days after liftoff. The mission would be tracked by stations in Germany, Santiago (Chile), Bermuda and Woomera. For the final stages of the descent, a mobile station with S- and C-band antenna was set up in Coober Pedy, Tjaliri, Australia.

Fitting out the *Express* took 22 months and in December 1994 it was shipped to Japan for integration with its *Mu* rocket. *Express* rode on top of the last *Mu-3SII* rocket to be fired from Kagoshima at 1.45 p.m. UTC on 15 February 1995, blazing into a night-time sky. An attitude control problem with the second stage 130 sec into the mission meant that control was lost for 20 sec and fuel was depleted as the rocket tried to compensate. The problem was later attributed to aerodynamic effects of a heavy payload during ascent (a similar problem caused the explosion of two Chinese rockets at around the same time).

The planned orbit of 210–398 km was not achieved. The resulting orbit was so low— possibly as low as 110 km—that it was considered that the cabin must burn up. United States Space Command never detected the launch nor entered it in its logs. However, the downlink signals confirmed that the Russian computer had stabilized the cabin and the mission has begun. Kagoshima established radio contact with the satellite at 3.30 p.m., Santiago at 4.01 p.m., Kagoshima again at 4.51 p.m.

Attempting to salvage something from the mission, German ground control in Oberpfaffenhofen, Munich, asked the tracking centre in Santiago, Chile, to command the satellite to reenter, hoping that they might at least recover the capsule rather than wait helplessly for it to be destroyed. When *Express* made its next pass over the centre at 5.30 p.m., Santiago received no signals and had no idea whether the reentry command had been received.

So the story ended. It was presumed that *Express* had burned up somewhere over the Pacific after about three orbits. The failure of the *Mu* was blamed on the heavy payload—twice that of the *Mu*'s previous heaviest assignment. Extra propellant had been loaded to all the *Mu* stages, computer simulations indicating that the profile would work. The computer must have been wrong.

Express

Several months later in Britain, Geoffrey Perry, the science teacher known for his role in listening to signals from early Soviet space probes, was alerted to a report in the *Ghanaian Times* of 3 February 1995, by-lined by Gariba Ibrahim in Tamale, which reported that a strange object with Russian markings on an orange parachute had descended from the skies at Kotorigu in west Mamprusi, near the border with Togo. The newspaper recorded the fact that the deputy commissioner of police, Patrick Agboda, had visited the area and noted that the bushes surrounding the fallen object were burned. Kotorigu is very rural, lacking electricity, water or telephone and the road can only be used by four-wheel-drives. The people living in the traditional African villages in the area must have been startled by the sonic boom which preceded the red-hot object descending under its parachute. The local chief, it was later learned, told people to keep away from the object from space. He ordered his brother to guard it while he went to the district chief, Mr Gumah of Walewale.

A photograph later appeared in the *Ghanaian Chronicle* on 20–23 February, by-lined by Abdulla Kassim in Tamale, which showed the strange object hanging from a tree and local people posing behind the spread out parachute. It was clear to Perry that it looked very like a FOBS reentry capsule and indeed the article quoted the commander of armed forces in the region, Group Captain Aryetey's conviction that it was Russian. Perry noted that the ground track of *Express* brought it over west Africa four hours after launch. Perry contacted the Ghanaian authorities a number of times in the course of the year to ascertain if the strange object was indeed the *Express* capsule, but fruitlessly. Having made no progress, he then sent an article on the missing capsule to the November 1995 monthly news bulletin of the Western Australian Astronautical Society with his speculation as to the true fate of *Express*.

In fact, more had gone on than he realized. District chief Gumah had organized a transport to take the *Express* away. Ten strong men had been required to put the capsule onto a truck. Hundreds came to look at the object in Walewale, where it had become something of a local attraction and it was handed over to the army. Two weeks later, the Ghanaian air force took the object to the town of Tamale, 50 km south, where it was stored in a huge aircraft hangar. This hangar had been built by the Russians in the 1960s during a period of intense Soviet–Ghana cooperation and the base commander, who had been there at the time, at once recognized the Cyrillic script on the parachute (lettering on the cabin itself had burned off).

Meanwhile, Perry's article in the news bulletin of the Western Australian Astronautical Society had been read by some members of the recovery team in Woomera, who must have been wondering what had happened to their seriously overdue cabin. They sent the article to DARA in Germany, who phoned the German embassy in Accra. After a few quick enquiries, it was ascertained that the strange object was almost certainly the *Express*.

In January 1996, officials from the German space agency DARA arrived in Ghana, identified the *Express* and asked for their satellite back. They found that the capsule had been moved to Tamale airport where it was lying, untouched and unopened, in a corner of the hanger. It was a little dented, not from its brief sojourn in space but from its truck ride along rutted African roads. Until the Germans arrived, no one had known quite what to do with it. The military authorities had set up a scientific commission to determine if it was a radioactive warhead (they drew a negative). They had called the Russian embassy in Accra, who quickly said it was not their satellite. The articles in the Ghanaian press about the strange object were well read in the small European community in Ghana, though the German members seem to have missed it and, when the cabin had been declared lost, no one had thought to alert Germany's wide-flung outposts throughout the world.

Luftwaffe squadron 62 duly sent a Transall plane out to Ghana which brought back the *Express* to Germany in a wooden container. Subsequent examination of the cabin found that *Express* had entered the atmosphere at 5.50 p.m. on that January day 80 km above the Earth. Barrelling nose first into the atmosphere, the cabin survived the intense heat. Sensing the onrush of air 6 km high, the barometric system commanded the parachutes to open. The experimental packages were in perfect order and the experiment to test ceramic materials on the nose cone was declared a complete, if belated, success!

MUSES-B: INTRODUCING THE NEW *MU-5* LAUNCHER

Muses-A, the first Moon flight, was the first of three Japanese spacecraft devoted to unconventional missions. The next, Muses-B, was an experiment in very long baseline interferometry and marked the first launch of the *Mu-5* launcher, successor to the *Mu-3*.

Mu-5 history	
12 Feb 1997	Muses-B/*Haruka*
4 Jul 1998	Planet B/*Nozomi*
10 Feb 2000	Astro-E (Fail)
2002	Muses-C (scheduled)

The objective of the *Mu-5* was to double the lifting power of the *Mu-3*. Approval for the new launcher had been given in 1989. Built by Nissan, *Mu-5* weighed 135 tonnes at liftoff, making it the largest ever *solid-only* propellant launcher ever flown. *Mu-5* is 2.5 m in diameter and can lift 1800 kg into Earth orbit. Originally, the division of responsibilities between ISAS and NASDA had limited ISAS to launchers of 1.4 m diameter or less, but this was at last waived.

The first stage, the M-14, used a mixed fuel of 68% ammonium perchlorate, 20% aluminium and 12% PBHT inside high-strength maraging steel. Efficiencies were achieved by using a collapsible nozzle on the first stage, extendible nozzles on the third and fourth stage and side jets. Another improvement was the use of fibre optical gyros to sense vehicle attitude. Development of the rocket was slow and by the time it was launched had fallen several years behind schedule, the main cause being the difficulty in developing retractable nozzles. *Mu-5* required the upgrading of the *Mu* pad at Kagoshima. Although much larger, the *Mu* still followed the same launch procedure as the older series, being tilted at an angle on a crane beside its launch tower. The cost of a launch was ¥13.20bn (€118m).

Shooting aloft on a pillar of bright yellow flames against a calm sea and clouds, the *Mu-5* placed the 830 kg Muses-B into an elliptical 573–21 402 km, 6.5 hr orbit on 12 February 1997. Once in orbit, it was renamed *Haruka*, meaning 'faraway'.

Haruka's basic shape was conventional enough—a box with solar panels. What made it unusual was the 8 m wide, wire-mesh, golden-coloured antenna which on the 17th day of the mission unfurled like a giant petal, topped by a reflector which peeped through the thin structure. *Haruka* worked on the principle that its radio telescope could be combined with another one on Earth to make a very long baseline, thus obtaining a wide measuring base and improving the resolution. It was designed to have a resolution of 90 microarcseconds, or a hundred times better than the Hubble Space Telescope (though in the radio, rather than the visual medium).

Stage	Engines	Thrust

Mu-5 (Nissan)

Length: 30.7 m
Weight: 135 tonnes
Diameter: 2.5 m
Performance: 1.8 tonnes to low Earth orbit; 800 kg to Sun synchronous orbit;
800 kg to geostationary transfer orbit; 520 kg to the Moon; 450 kg to Mars

Stage	Engines	Thrust
1	M-14 solid	395 tonnes
2	2413 solid	140 tonnes
3	M-34 solid	30 tonnes

The ¥8,664m (€77.3m) *Haruka* mission was developed by ISAS in cooperation with colleagues in the United States, Canada, Australia and Europe. *Haruka* was tracked by stations in five locations—Usuda, Greenbank, Goldstone, Madrid and Tindbinbilla, where it sent data at 128 megabytes a second. It was designed to record active radio sources in the universe, in galaxies, quasars, pulsars and masers. One of it first observations was of quasar 1156+295. Whereas ground pictures showed a reddish-yellow blob, Haruka images were much more precise, showing clearly a jet of material being ejected from the core.

NOZOMI TO MARS

The Planet A mission to Comet Halley was, as the name suggested, the first Japanese planetary-type mission. The Planet B mission was defined as Japan's first venture to a terrestrial-type planet and was originally a plan to launch a small payload to Venus, the nearest planet and the easiest to reach. The objective changed when it became clear that Mars, although further away, was scientifically more promising.

Planet B became Japan's first spacecraft to Mars and was launched on 4 July 1998. Soon after its launch, it was renamed *Nozomi*, or 'hope'. The objective was to orbit Mars at an altitude of 300 to 47 500 km, the low point later adjusted to 130–150 km and transmit data for one Martian year. Fourteen instruments, weighing 33 kg, were designed to send back information on the magnetic field, on the structure of the atmosphere (turbulence, motion and seasonal variation) and on energetic particles, with a particular view to answering the question as to why Mars lost its water. The scientific instruments came not just from Japan but also from France, the European Space Agency, the Swedish Institute of Space Physics, the Canadian Space Agency, NASA and Munich Technical University. NASA had a neutral gas mass spectrometer to measure the chemical composition of the Martian atmosphere. A Japanese/French camera was carried. Several passes were to be made of Mars' tiny moons, Phobos and Deimos.

The project cost ¥18.6bn yen (€166m). Although the start of the mission attracted little attention abroad, there was much more excitement at home, where 270 000 people

Mu-5 launching

had their names inscribed in miniature on the spacecraft. Its weight was 540 kg, making it one of the smallest spacecraft to travel to Mars.

Nozomi was a hexagonal box shape with a bell-shaped motor underneath and a dish antenna on top. The spacecraft measured 1.6 by 1.6 by 0.6 m. The box comprised solar cells and experimentation. Overall, the spacecraft was taller than the Japanese technicians who fussed over its final stage of assembly in a clean room and ladders had to be used to work on the top part of the probe. *Nozomi* had a 500-kg thrust engine to carry out a complex set of manoeuvres, first in Earth orbit, then in trans-Mars coast, and then to subsequently enter orbit around the red planet. Once away from Earth, thin, 25 m long radio wire antennas were deployed. Also sticking out was the 5 m magnetometer.

Nozomi over Mars

> ***Nozomi's* instruments**
> Camera
> Magnetometer
> Instrument to measure energetic electrons
> Thermal ion drift measurer
> Electron spectrometer
> Instruments to measure energetic ions and their mass
> Plasma wave detector
> Neutral gas mass spectrometer
> Dust counter
> Extreme ultraviolet spectrometer
> Ultrastable oscillator for radio science
> Instrument to measure high-energy particles
> Wave sounder

Twenty-three minutes after liftoff, the spacecraft entered a swinging Earth–Moon orbit of 703 to 489 382 km. Its solar panels sprung open. At a far point out in the trajectory, 300 000 km from its home planet, its cameras were turned on to capture an unusual view: the Earth–Moon system in the same frame. A thin crescent of Earth filled the bottom left of the picture, while in the far right could be made out the crescent of Earth's Moon. On

24 September, as it came near to the Moon for the first time, it used lunar gravity to swing the spacecraft into an even more extreme orbit, one as far out as 1.7 million kilometres. Cameras clicked to return spectacular pictures of the far side of the Moon, picking out the Moscow Sea and the pitch black floor of crater Tsiolkovsky. This was in preparation for a manoeuvre on 20 December to nudge it out of the Earth–Moon system and on its way to Mars.

Nozomi flew around the Moon at a distance of 2809 km on 18 December and then, firing its little engine, passed 1003 km from Earth two day later, but the burn at perigee proved insufficient to kick the spacecraft onto a trans-Mars path. It later transpired that the valve supplying oxidizer had failed to open sufficiently, wasting fuel and not producing sufficient thrust either. As a result, a second burn had to be made. This failed to rectify the situation, for it turned out that the fuel supply had been severely depleted at this stage. However, the burn did manage to place *Nozomi* on a slow solar orbit, one which would involve three orbits of the Sun, a swing by Earth in December 2002 and June 2003, sufficient to eventually reach Mars sometime in either December 2003 or January 2004, four years late. Ground controllers were confident that the spacecraft's equipment would remain in sufficiently good condition to carry out its delayed mission in full.

CONCLUSIONS

During the 1980s and 1990s, Japan used its small solid fuel rockets developed by ISAS to launch a series of scientific missions. These ranged from an important set of scientific missions dedicated to studying the Sun and stellar objects to more adventurous missions to Comet Halley, the Moon and Mars. Long before the bigger space powers began to preach the virtues of smaller, cheaper spacecraft, Japan launched tiny Planet and Muses missions on small rockets, showing what could be done by careful and imaginative planning, budgeting and miniaturization.

4

Advances in technology, engineering and applications

In the 1970s and 1980s, using American-based rockets, the N-I, N-II and H-I, the Japanese laid the groundwork for a modern space programme. The 1990s saw the introduction by NASDA of a powerful domestic-built rocket, the H-II and significant advances in technology, engineering and applications.

H-II ROCKET: 'MOST ADVANCED OF ITS KIND'

The H-II was in effect NASDA's declaration of launching independence, for the H-II used almost entirely indigenously produced technology, unlike the bought-in designs of the N-I, N-II and H-I. Approval was given for the project in August 1986. The H-II could thus legitimately be offered on the open world launcher market without breaching licensing arrangements. To do this, a complex structure was built up in which NASDA funded the H-II's development, the rocket was built and marketed by the newly created Rocket Systems Corporation for a fee, the company then selling it abroad if possible, or, in the domestic market, selling it back to NASDA for use on a mission-by-mission basis. Rocket Systems Corporation included many staff formerly with NASDA.

It was also madly expensive, one of the consequences of the Japanese drive for quality assurance and elaborate testing. The engineers did not want the H-II to be first Japanese rocket to explode. They were later criticized for their over-indulgence in using expensive materials and backup systems. However, the Japanese rocketeers also suffered from the smaller size of the Japanese market which meant that economies of scale available in the United States were not possible in Japan.

With the H-II, Japan was able to send 10 tonnes into low Earth orbit or 4 tonnes to geostationary transfer orbit. The latter figure was better than China's *Long March 2E* (3.4 tonnes), the American *Delta 3* or *Atlas 2AS* (3.8 tonnes), but inferior to the European *Ariane 4* or *5* (4.5 and 6.8 tonnes respectively) or Russia's *Proton* (4.8 tonnes).

The aim of the H-II rocket was to send payloads to geostationary orbit and be the mainstay of Japanese rocket launches in the 1990s. The LE-5 engine, used on the second stage of the H-I, was used again as the second stage of the H-II, but with extra thrust. The

H-II (Mitsubishi)

Length: 49 m
Diameter: 13.1 m
Weight: 266 tonnes
Capacity: 2 tonnes GEO; 4 tonnes GTO; 10 tonnes LEO

Stage	Engines	Thrust, kg
Strap-on	Two SRM solids	$2 \times 158\,760$
1	LE-7 liquid hydrogen	85 730
2	LE-5A liquid hydrogen	12 247

Bird's-eye view of H-II at liftoff

LE-5A, as it was called, could be restarted and tilted for pitch and yaw. But the big difference was the use of a liquid hydrogen first stage with a new LE-7 high-thrust, high-temperature high-performance engine able to generate nearly a hundred tonnes of thrust, in design since 1987. Development of the engine proved problematical, a series of fires, explosions, welding problems and mechanical failures putting the programme two years behind schedule.

The building of the cryogenic fuel tanks posed special engineering challenges. The tanks were built by Mitsubishi with the assistance of the Sciaky Corporation of Chicago and used the same alloy as that employed for the American space shuttle, aluminium 2219. Temperatures were kept down by foam-resin insulation.

Additional thrust was provided by two large solid rocket boosters, almost as large as those of the American *Titan IV* and the space shuttle. The nozzles on the solid rocket boosters could be swivelled 5°. The H-II was imposing in size, comparable to the Chinese *Long March 2E* and Europe's *Ariane 4*. The authoritative American journal *Aviation Week & Space Technology*, described it as 'the most advanced expendable launch vehicles sitting on the pad today in terms of its integration of modern materials, electronics, computers and propulsion' [3]. When the Americans came to construct their *Delta 3* in the mid-1990s, they used fuel tanks made in Japan. Technology transfer between Japan and the United States was now flowing the other way.

For all its innovation, the H-II had an unhappy development history. The schedule got behindhand from an early stage. In 1989, a bad mixture of liquid oxygen and hydrogen caused an engine fire which destroyed the test stand for the engine. A fire broke out during one of the early test firings of the new LE-7 engine in July 1990, melting piping and pumps. In April 1991, four test firings of the first stage engine failed. On the last one, the engine shut down 14 sec into a 350 sec firing, having failed to reach adequate combustion pressure. On 9 August 1991, there was an explosion in the Nagoya Guidance and Propulsion Systems works in Komaki, 200 km west of Tokyo. During the pressurization test of the LE-7 to 143 atmospheres, an engine valve burst causing an explosion and the death of a technician. The main problem appeared to be the LE-7 cryogenic turbopumps whose job was to supply high-pressure propellants to the combustion chamber. In 1992, the LE-7 failed five seconds into a major engineering test and caught fire because of a fuel leak. Then in June 1992, during a 10-sec firing of the LE-7, fire broke out five seconds after ignition when a weld broke and permitted liquid hydrogen to escape. Although the fire was extinguished in a minute, it was so intense that the engine broke away from its mounting and fell 23 m into the cooling pond! The hydrogen turbopump was also found in the water. The following year, the LE-7 aborted a test firing on the Yoshinuba stand 132 sec into a 350-sec run, not operating smoothly until the summer.

SOUNDING ROCKET PRECURSORS

Typical of Japanese thoroughness was the use of a sounding rocket to test elements of the H-II design. The TR-1 series was a sounding rocket designed to pave the way for the H-II rocket. The TR-1 was a quarter-scale model of the H-II, though it had to be equipped with large rocket fins which were unnecessary for the H-II itself. These rockets were

14.3 m long, 1.1 m in diameter and weighed 11.9 tonnes. The purpose of the launchings was to obtain important data on rocket stresses, pressures, heating and exhaust plumes, as well as to test the mechanisms for the release of the strap-on boosters. The first TR-1 test was made on 6 September 1988. A second mission, using dummy solid rocket booster strap-ons, was launched 27 January 1989 and recovered from the sea several hours later. The programme concluded with a third launch on 20 August 1990. These rehearsals provided valuable testing information—but probably pushed costs up.

Ryusei shooting star

SHOOTING STAR

The H-II eventually went for its first real test on 4 February 1994. As if to lay to rest the ghost of its unhappy history behind, H-II rose smoothly from the ultramodern Yoshinuba complex of the Tanegashima launch centre on an ambitious space experiment. The twin solid rocket boosters dropped off 1 min 34 sec into the mission at an altitude of 37 km. The first-stage engines burned out at 6 min when the H-II was 227 km high and 630 km downrange by which stage the vehicle was nearing Christmas Island. The second stage burned for five minutes. Executive director Tomihumi Godai was still in the control room when a congratulatory phone call came through from the Prime Minister, Mohiro Hosokawa.

Fourteen minutes into the mission the H-II released OREX, the Orbital Reentry Experiment. This was a flattish flying-saucer-shaped dome, 3.4 m in diameter and 1.46 m high weighing 865 kg designed to test the materials which one day might be used on a Japanese spaceplane. The materials of OREX were a mixture of carbon and ceramic tiles. OREX, renamed *Ryusei* or 'shooting star' once in space, made a single orbit around the Earth. Following a five-minute reentry burn, thrusters adjusted the angle of reentry where the surface temperature glowed white-hot to a temperature of 1570°C, hotter than reentry on the space shuttle. The shooting star blazed through the atmosphere over the Pacific, came through radio black-out and splashed down 460 km south of Christmas Island, though the payload was not recovered.

Meanwhile, the H-II had placed in orbit a 2.4-tonne, prosaically termed 'vehicle evaluation payload', more poetically *Myojo*. It role was to enter space to monitor the rocket's performance, acceleration and vibration and circle the Earth in a high-altitude trajectory for four days. *Myojo* was a box-shaped object carrying 1.5 tonnes of water which it pressurized, depressurized and repressurized in a series of tests.

For Japan, the successful launch was a great moment. 'First large launch vehicle using only domestic technologies' trumpeted the media' [4]. The H-II rivalled the best of what the Americans, the Russians and the Europeans could offer and it had all been developed at home this time.

The second H-II launching was a disappointment. The payload was the sixth engineering technology satellite, ETS-6 or *Kiku-6*, which was intended to test lasers (the first such Japanese tests), advanced telecommunications systems and new propulsion methods paving the way for data relay satellites and operational ion engines. The ion engine was rated at 23 mN and was designed to work for 6500 hours. On the first launch attempt, the rocket engines failed to fire. When the H-II did get off the pad on 28 August 1994, the ¥52.75bn (€471m) 1.6-tonne satellite was left in an initial 236–36,338 km orbit instead of geosynchronous one. The apogee burn had generated only 10% of the thrust needed. Despite the setback, *Kiku-6* was able to test its ion engine, new nickel hydrogen batteries and an electro-thermal-hydrazine thrust. The intersatellite communications systems were tested out. The ion engine reached the designed level of thrust (23 mN). The remaining thrust was used to raise the perigee to 8560 km by the end of November. Because the orbit passed through the Van Allen radiation belts, the solar cells and electrical equipment degraded quite rapidly. Making a virtue of necessity, the opportunity was used to relay information on electrons and protons in the radiation belts. The mission ended in January 1996.

H-II history	
4 Feb 1994	OREX/VEP
28 Aug 1994	ETS-6/*Kiku-6*
18 Mar 1995	SFU, GMS-5/*Himarawari-5*
17 Aug 1996	ADEOS-1/*Midori*
28 Nov 1997	TRMM/ETS-7/*Hikoboshi/Orihime*
21 Feb 1998	COMETS/*Kakehashi*
15 Nov 1999	MT-SAT (failure)

H-II BRINGS IN ERA OF ILL-LUCK, UNCERTAINTY

The *Kiku-6* failure began a run of ill-luck which dogged the Japanese space programme for the rest of the 1990s. Two years later, the HYFLEX mini-shuttle was lost at sea after a sub-orbital mission (though on a different launcher). The next year saw the loss of the ADEOS 1 satellite and in 1998 the placing of another, COMETS in a virtually useless orbit. These four losses were valued at €461m, €42m, €943m and €353m respectively (total: €1.809bn (¥202.6bn)). In 1997, the Science and Technology Agency publicly criticized the use of large, expensive satellites where a single fault could jeopardize a multi-billion yen project, arguing instead for smaller, cheaper, less ambitious projects where a single failure would prove less costly.

In a programme where quality control was emphasized throughout, this was an alarming series of events and the government established a high-level panel to examine the problems. Formed of eight Japanese and eight foreigners, the panel was headed up by Jacques-Louis Lions, president of the French Academy of Sciences, and Tokyo University professor, Jiro Kondo. The panel reported in 1999.

The panel was unable to find any common thread in these failure (a conclusion common to similar investigations of a run of space disasters in other countries at the same time). The panel had many positive comments to make on NASDA's programmes, such as the performance of the H-II and the advanced technology used on Japanese communications and meteorological satellites. This had been achieved with limited resources. The panel noted that although NASDA's annual spending had risen from ¥50bn in the 1970s to ¥100bn in the 1980s and ¥200bn in the 1990s, it had declined as a proportion of Japan's gross domestic product from 0.04% to 0.035%. Staffing had not increased since the 1980s, even though the responsibilities and commitments of the agency had grown. NASDA's budget was less than half that of Europe's and less than a tenth that of NASA. Much had been achieved with little.

Critically, the panel found poor lines of management communication, a lack of clear responsibilities for project management, the need for more technical resources in some projects and the need for improved systems of quality assurance. It criticized NASDA for taking on too many projects, for failing to concentrate its resources and for being the junior partner in too many international projects. The programme was failing to generate commercial benefits at home: although Japan had led many of the new communications technologies, Japanese companies still preferred to buy in foreign communications satellites.

All this came at a bad time. The financial crisis which hit the far east in the mid-1990s took its toll on the Japanese space programme. In 1997, the Space Activities Commission agreed, following pressure from government, to cut space spending by 14%, to be achieved by schedule slippages, the importation of foreign technology and the redesign of several programmes, like the H-II and the *HOPE* spaceplane.

To cap the tale of woe, one of the last H-IIs veered off course on 15 November 1999 and had to be destroyed—Japan's first launch failure since 1970. MT-SAT was a project of the Ministry of Transport and was the first major satellite project conducted by a body other than ISAS or NASDA. Built by Loral, MT-SAT stood for Multifunctional Transport Satellite. It was a box-shaped 4 by 4 m, 1600-kg satellite designed to provide weather observation and air-traffic control, navigation and communications from 24-hr orbit. MT-SAT was to replace the last of the GMS series, *Himawari-5*. MT-SAT had four infrared sensors (one with a 1-km resolution) and one in the visible channel to provide weather data on cloud height and distribution, wind conditions and temperatures. The ground-breaking aspect of MT-SAT was that it was intended to anticipate the continued rapid growth in air travel over the north Pacific region by ensuring a much higher level of air-traffic control safety through improved communications and navigation. Its 1999 launch

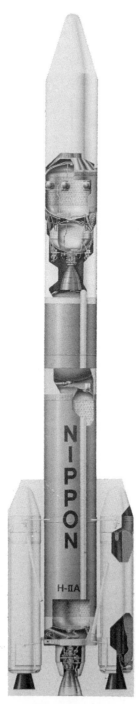

H-IIA: cutaway view

was postponed by a series of problems in the H-II rocket, such as a leak detector failure and electronic malfunctions. When it did eventually take off, the main engine cut off late, the rocket began to veer off course, telemetry was lost and the destruct command was reluctantly sent in at 7 min 41 sec. For Japan, it was a sad and undeserved end to a troubled decade. It was decided to launch no more H-IIs.

AUGMENTED: H-IIA

The H-II, although a great technological achievement, was much more expensive than ever anticipated. When the H-II first flew, the Japanese were confident that satellite companies would pay the extra money involved in exchange for its reliability. Japan entered the H-II in a number of international competitions in the early to mid-1990s, to be repeatedly beaten by Europe's *Ariane* and the Chinese *Long March*. The financial pressure became even more acute when Russia's *Proton* entered the world launcher market, with early success. New American launchers arrived.

It was decided to phase out the H-II in the late 1990s, to replace it with a more reliable, modernized, economical version, the H-IIA, 'A' standing for 'augmented'. Each H-II launch cost around €188m, twice that of an equivalent launch by *Atlas* or *Ariane* and it was NASDA's objective to bring the cost of the H-IIA launch down to between €94m and €141m (¥10.5bn to ¥15.7bn), though a more ambitious figure of €80m (¥8.9bn) was later set. To do this, NASDA decided, reluctantly, to use imported technologies, with American-based solid rocket boosters, able to offer 10% more power. New solid rocket engines were selected, to be supplied by the American Thiokol corporation, the company which built the boosters for the space shuttle. The redesign involved a new engine, the LE-7A, a very much simplified version of the H-II's LE-7, with new cryogenic tanks, redesigned plumbing, pre-burner chamber and turbopump systems, able to offer 13% more power. The development programme for the new rocket was set to cost €696m (¥77bn). Other economies were made: the weight of the second-stage tank was reduced, copper tubing replaced stainless steel tubing, the amount of engine welding was reduced and production time for the engines was halved. Typical of the economies

H-IIA (Rocket Systems Corp)		
Length: 52 m Diameter: 4 m Weight: 389 tonnes to 410 tonnes (according to version) Performance: 3.3 tonnes to geostationary orbit		
Stage	Engines	Thrust, kg
Strap-ons	2 SRM solids	2 × 190 058
1	LE-7A, liquid hydrogen	85 730
2	LE-5B, liquid hydrogen	14 061

made by the H-IIA, the solid rockets were shorter and fatter than those of the H-II, comprising a single segment rather than four, all of which had previously to be stacked separately, an expensive and time-consuming process. Whereas the performance of the H-IIA would not be dramatically different from the H-II, the price difference would be substantial. Japan had the ambition that the H-IIA gain up to 17% of the world launcher market by 2003 [5]. There was high confidence that the H-IIA would succeed and a first production run of 23 rockets was ordered.

Five versions of the H-IIA were announced, as follows:

Version	Features	Performance
H-IIA-202 (basic version)	Two solid rocket boosters	4.5 tonnes GTO 2.2 tonnes GEO 10 tonnes LEO 4 tonnes to SSO 2.9 tonnes to TLI
H-IIA-2022	Two solid rocket boosters and four strap-on boosters	4.25 tonnes to GTO
H-IIA-2024	Two solid rocket boosters and four strap-on boosters	4.5 tonnes to GTO
H-IIA-212 (for HOPE Transfer Vehicle)	Two solid rocket boosters and one liquid rocket booster	6 tonnes to GTO 3.3 tonnes to GEO 14 tonnes into LEO 5 tonnes to SSO 5.4 tonnes TLI
H-IIA-222	Two solid rocket boosters and two liquid rocket boosters	9.5 tonnes to GTO 23 tonnes to LEO 6.6 tonnes to TLI

LEO = Low Earth Orbit; GTO = Geostationary Transfer Orbit; GEO = Geostationary Earth Orbit; SSO = Sun Synchronous Orbit; TLI = Trans Lunar Injection

The most unusual variant is the H-IIA-212, designed for the HOPE Transfer Vehicle, a supply ship for the International Space Station. The core will carry a single liquid-fuelled strap-on at its side, one as large as the first stage itself, giving it a lopsided appearance.

The first new solid motors for the H-IIA were successfully tested in 1998. Development of the new LE-7A engine proved to be more problematical, principally with the wiring systems and nozzle vibration. The third test failed due to a hairline crack while another crack caused the fifth test to be called off 3 sec into a 350 sec firing. There was a further premature shutdown in a June 1999 test. However, long-duration engine tests were successfully completed in September 1999, putting the programme back on course.

Tests of the new second-stage engine were more encouraging. The LE-5B was designed to achieve 14 000 kg of thrust compared to 12 500 for the LE-5, but to be lighter at the same time. Thirteen successful firings were carried out by Mitsubishi at its Akita, northern Japan test stand in 1996, including continuous firings of more than 300 sec. In order to prepare for its introduction on the H-IIA, it was decided to give the LE-5B a live test on the H-II on the MT-SAT launch in November 1999, leading to the launch being renamed 'the half H-IIA'.

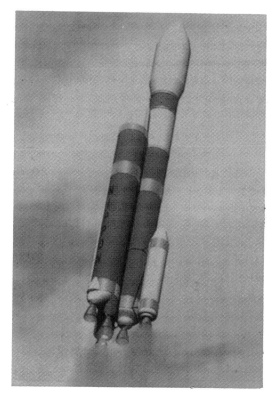

H-IIA—212 version

NEW SMALL LAUNCHER: THE J-1

The tribulation affecting the Japanese space programme resulting from the problems of the H-II was exacerbated by similar problems associated with the introduction of another, smaller launcher, Like the H-II, its early development was marked by setbacks and problems of cost.

In April 1993, Japan began planning the J-1 launch rocket, designed to combine the experiences of ISAS and NASDA. The idea was to develop a low-cost rocket able to send up payloads of up to a tonne. The J-1 used the solid-rocket strap-ons from the H-II as its first stage and the second and third stages were adapted from the old *Mu-SII* second and third stages. Again, to make further economies, it used the old Osaki launch pad at Tanegashima.

J-1 history
11 Feb 1996	HYFLEX
2001	OICETS (scheduled)

J-1 (Mitsubishi)

Length: 33.1 m
Diameter: 1.81 m
Weight: 87.7 tonnes
Performance: 870 kg into circular 260 km orbit, 30°;
1000 kg to LEO

Stage	Engines	Thrust, kg
1	SRM solid	181 440
2	M-23 solid	42 048
3	M3B solid	11 838

J-1 ready for night launch

J-1 flew only the first two stages on its inaugural flight on 12 February 1996. It was an up-and-down sub-orbital mission. The long, thin rocket, decorated with red stripes and the word 'Nippon' running down the side, took off from the old N rocket pad. The main payload was the HYFLEX spaceplane, a one-seventh model of the planned HOPE automatic space shuttle. HYFLEX was a snub-nosed spaceplane with stubby tail wings made out of aluminium and ceramic tiles. The J-1 sent the 1048 kg, 4 m long lifting body through the full profile of mach 14 during a 7.6 min mission over 100 km high. Released at 3.9 km/sec at an altitude of 110 km, HYFLEX came through sub-orbital reentry, glowing red-hot at a 47° angle. After passing through reentry, HYFLEX dived nose-down at 30°, making a series of banking manoeuvres, until the parachute opened 300 km northeast of Chichinjima, near the Bonin Islands. All went well until splashdown when the flotation harness broke free of the spaceplane which sank in 4000 m of water, too deep to be salvaged, 200 km northeast of the Bonin Islands. Of the 14 key areas of data required, NASDA obtained information on 12 by radiotelemetry, but missed out on two because of the sinking.

This was an unfortunate end to an otherwise successful debut. The J-1 proved to be much more expensive than hoped or planned, coming in at about ¥5.283bn (€47m) a launch, which was about double the price that would be competitive for a small launcher. A report on the rocket in April 1998 by the General Affairs Agency criticized not only the rocket project but also NASDA for mismanagement and inefficiency. The following month, the National Management & Coordination Agency, a governmental watchdog body, went further and suggested that unless costs of the J-1 could be lopped, then the programme should be dropped. The rocket had been supposed to save costs by the use of existing hardware but instead cost two to four times more than equivalent foreign vehicles and had only one payload customer. The agency took the opportunity to attack the costs of the *Mu-5* and suggested that the booster it had replaced, the *Mu-3S*, should be brought out of retirement.

As with the H-II, a redesign was ordered (cost: ¥6.49m (€58m)), one which would cut the cost in half, called the J-1 (the term J-2 has also been used). The plan involved:

- switch from solid fuel to liquid fuel to reduce costs;
- use of foreign suppliers to cut costs from ¥4.435bn (€39.6m) to ¥2.9bn (€36.4m) a launch;
- use of Russian NK-33 first-stage rocket motors;
- second stage to be powered by methane natural gas supplied by Ishakawajima-Harima Heavy Industries using a cryogenic carbon-fibre second stage tank to save weight;
- new dimensions: length 37.9 m, diameter 3 m, thrust 180 tonnes, first launch in 2002.

The NK-33 engine is named after Nikolai Kuznetsov and was developed as the second-stage engine of the Russian manned Moon rocket of the 1960s. When the programme was cancelled in 1976, all the equipment related to the project was ordered to be destroyed. Instead, Kuznetsov and his colleagues at their factory in Samara hid the engines away, awaiting better times. In the 1990s, they were discovered by the Americans and the Aerojet company bought them with a production licence. They found, as did other American companies, that old Russian rocket engines were powerful and reliable, offering them quicker access to space than they could develop themselves. However, a further review was ordered in 2000.

COMMUNICATIONS: LEADING THE REVOLUTION

The Japanese established their first operational satellite systems in the late 1970s and built an extensive system. By the late 1980s, communication satellite launches had become almost routine, using both Japanese and foreign launchers. Until the mid-1990s, Japanese comsats had served largely a domestic market.

1989 was a watershed year in the development of Japanese communications. That year, the Japanese parliament, the Diet, lifted restrictions on commercial development, leading to the formation of new companies, alliances and consortia.

The Ministry of Posts and Telecommunications had led the development of Japanese communications satellites in the 1970s and 1980s. NASDA was not permitted under its charter to handle commercial activities, so telecommunications and other revenue-generating space activities were contracted out to government–private companies or private consortia linking banks, insurers and companies active in space-related businesses. For government, this had the advantage of spreading risks, involving industry in the planning process, ensuring the product could be of maximum benefit to industry, but the indirect effect of creating such close government–industry ties as to make the market impenetrable to foreign companies, especially the Americans. In a tangled set of institutional arrangements, the ministry contracted out the JCSat and *Superbird* programmes from the state communications monopoly to the Telecommunications Advancement Organization of Japan which procured NASDA to launch and operate the satellites.

Under section 301 of the 1974 United States Trade Act, America eventually took action against countries engaging in what it deemed to be unfair trade practices, such as preventing access to domestic markets or over-subsidizing their own industries. Japan was the top target, satellite markets being one of the main categories within the objections. The Americans complained that these arrangements froze out foreign competition, although American communications companies had long been sub-contractors to the large Japanese telecommunications companies. In the course of the Japanese–American trade disputes of the 1980s, they put pressure on Japan to deregulate the system. The market was duly liberalized by parliament in 1989 and by the US–Japan Trade Impediment Initiatives Agreement of 1990. Its first test was the new N-Star generation of communication satellites.

The N-star series was essentially the CS-4 or *Sakura-4* series. It was originally pro-moted by NASDA as an experimental series and, as such, closed to foreign competition. Following American pressure, the experimental aspects were transferred to the COMETS programme while the CS-4 satellites, renamed N-Star, went for open competition, the American Loral and Hughes companies becoming satellite contractors and *Ariane* win-ning the launch contract (August 1995 and February 1996). Each weighed four tonnes and had 26 transponders.

The early 1990s saw the Japanese communications industry hit hard by launch failures. Two Japanese communications satellites—one was a BS-2X direct broadcasting

Sakura communications satellite

satellite, the other *Superbird B*—were destroyed when *Ariane* V36 exploded at Korou in French Guiana on 22 February 1990 6 sec after liftoff. Recovery of wreckage from mangrove swamps determined the cause as a dirty rag left in a tubing pipe supplying water to the engines.

The BS satellites were built for the Japanese Satellite Broadcasting Company and NHK to provide three television channels to 300 000 subscribers in Japan. The loss jeopardized the company's ability to provide a service to them. Later that year, BS-3A/ *Yuri-3A* was launched on H-1 in August 1990 but a quarter of its electrical power was lost because of a short circuit in one of the solar panels, a problem which worsened when the sun angle reduced electrical power still further.

After the *Ariane* failure, the replacement satellite, the backup comsat, BS-3H was reallocated to an American *Atlas* launcher. In the event, the backup was lost too when during its launch on 18 April 1991, one of the two *Centaur* upper stage engines completely failed to fire 4 min 44 sec into the flight. The *Centaur* cartwheeled around the sky, eventually exploding 77 sec later, 175 km high and 390 km downrange. Radar tracked débris raining down into the Atlantic for minutes afterwards.

The second backup satellite was eventually launched by Japan itself into cloudy skies on 25 August 1991. The BS-3B, also called *Yuri-3B*, carried three television transponders to send television out to 1.5 million homes. BS-3N was launched by *Ariane* on 9 July 1994. It was replaced by the first of the BS-4 series, *Yuri-4*, in 1997, the second being due in 2000 and intended to carry digital television.

Superbird B was launched on 26 February 1992 and *Superbird A-1* on 1 December 1992 on Ariane from Korou. *Superbird A-1* was 1665 kg in weight, box-shaped, 3.4 m high, 2.4 m by 2.2 m, with a 20.3 m solar panel able to provide domestic communications for the Japanese archipelago. *Superbird* had a beam able to send television signals to Korea, Australia, New Zealand, Hawaii, Malaysia and China. About 10 million people were receiving pictures from BS satellites by late 1999. JCSats were the first to serve the international Asian communications market, followed by *Superbird C* in 1997.

Japan's use of foreign launchers to put up communications satellites was never more evident than on the 29 August 1995 when, on the same day, an *Ariane* launched N-star

JCSat (Japanese Satellite)

1	6 Mar 1989	*Ariane*
2	1 Jan 1990	*Titan 3*
3	29 Aug 1995	*Atlas*
4	3 Dec 1997	*Ariane*

Superbird (Space Communications Corporation)

A	5 Jun 1989	*Ariane*
B	22 Feb 1990	*Ariane*—failure
B	26 Feb 1992	*Ariane*
A1	1 Dec 1992	*Ariane*

Communication Satellite, CS/*Sakura* (Nippon Telegraph & Telephone)

1	15 Dec 1977	*Delta*
2A	4 Feb 1983	N-II
2B	5 Aug 1983	N-II
3A	16 Feb 1988	H-I
3B	16 Sep 1988	H-I
N-Star A	29 Aug 1995	*Ariane*

BS/*Yuri* (NHK)

1	7 Apr 1978	*Delta*
2A	23 Jan 1984	N-II
2B	12 Feb 1986	N-II
2X	22 Feb 1990	*Ariane*—failure
3A	25 Aug 1990	H-I
3B	25 Aug 1991	H-I
3N	9 Jul 1994	*Ariane*

from Korou for Nippon Telegraph and Telephone and an *Atlas 2AS* fired aloft JCSat 3 for Japan Satellite Inc. A second N-Star was launched by *Ariane* in February 1996. JCSat 4, with 32 transponders to provide communications for Japan, Asia and Hawaii for 12 years, was launched by *Ariane* on 3 December 1997 and B-Sat 1B also by *Ariane* on 29 April 1998.

By the late 1990s, CS, JCSat, *Superbird* and their services between them provided 300 channels to Japan and had as many as 1.4 million receiving digital television. The world communications satellite market continued to grow in the 1990s, traditional geosynchronous satellites being joined by constellations of comsats flying in low Earth orbit, like *Globalstar*, designed for the global mobile market. A late 1990s satellite forecast anticipated the need for more than a hundred new comsats every year for 1999–2002, rising to 170 in 2003 [6]. Within Japan, several systems were under study in the late 1990s—low Earth orbit systems, high-rate data satellites and high elevation synchronous satellites able to send quality broadcasting to high latitudes.

EARTH AND MARINE OBSERVATIONS

Japan's first satellites were for scientific (ISAS), communications and engineering purposes (NASDA). During the 1970s, an awareness grew of the potential of space observation platforms to watch the seas and land masses for remote sensing. Satellites could be used to make maps, spot pollution, find fish, assess crops and study water resources. Application of imaging data from satellites could be used in thousands of ways to assist agricultural, marine and economic development. The United States launched the world's first remote sensing satellite, *Landsat* in 1972. The Russian series was called *Resurs* (Resource) and the French one SPOT (Satellite pour l'Observation de la Terre).

Responding to these developments, in July 1975 Japan established a Remote Sensing Technology Centre, Restec to lead the country's endeavours in Earth resources work. Three years later, Japan built an Earth Observation Centre at Hatoyama, Saitama, to receive *Landsat* data and started to take these data from 1979. In 1978, the Space Activities Commission took the decision that Japan should develop its own land and sea observation satellites and move away from dependence on American data. Preliminary designs were carried out in 1979–80 and the final design was settled in 1981.

MOMO—FOLLOWING THE TYPHOONS

The first Japanese remote sensing satellite was the Marine Observation Satellite, MOS, also called MOS-1A. Weighing 750 kg, it was designed to carry four instruments to study the surface of the Earth's oceans. Once in orbit, it was renamed *Momo*, or 'peach blossom'. It was launched into a 903–917 km orbit, 99.1°, period 103 min on 19 February 1987 on the last N-II rocket. This orbit was one which made 14 revolutions a day, repeating the ground track every 17 days. The prime contractor was NEC.

MOS-1A was a box-shaped satellite with one solar panel. It carried a 70-kg multi-spectral electronic self-scanning radiometer using a charge-coupled device, a 54-kg microwave scanning radiometer to observe temperature and water vapour and a 25-kg visible and infrared thermal radiometer. The scanning radiometer could compile a coloured sea map which could indicate pollution in fish-rich zones. *Momo* was able to measure atmospheric water vapour, ice floes, plankton, ocean currents and sea temperature. A gas jet system was devised to maintain station-keeping on orbit. A data collection system was installed in order to collect information from automatic monitoring systems and relay them back to ground control. As it crossed the oceans, its multispectrum radiometer covered a swath of 100 km, its microwave scanning radiometer 317 km and the visible and infrared radiometer 1500 km.

Tracking was done by Tsukuba Ground Centre, with support from other tracking stations at Katsura, Masuda and Okinawa. Under cooperative agreements, MOS data were received by Thailand (Bangkok), the National Institute of Polar Research in Showa Base, Alice Springs, in Australia and the Canadian Centre for Remote Sensing (Prince Albert and Gatineau).

The satellite worked for two and a half years until June 1989. It sent back outstanding pictures of tropical storms. Using false colour imaging, the swirl of a typhoon could be made out in dark blue colours, surrounded by red sea, the red colour being due to the warm temperatures from which typhoons sucked up their moisture. And, right in the centre of the typhoon, a small, 4 km wide red spot—the eye of the hurricane. Other pictures from *Momo* showed the snows on the top of mountains, floating ice around the northern Japanese islands, a volcanic plume streaming from Mount Fuji. Photographs of the Antarctic ice sheet were relayed to the Japanese observation party at Showa base there.

MOS/*Momo* series		
MOS-1A	18 Feb 1987	N-II
MOS-1B	7 Feb 1990	H-I

Three years later, the second marine observation satellite, MOS-1B or *Momo-1B*, was launched by H-I from Tanegashima on 7 February 1990, rising on a pillar of smoke into pacific clouds as waves lapped the island launchpad. The satellite was actually the backup model for MOS-1A. Also launched into virtually identical orbit with MOS were *Orizuru* (meaning 'beginning'), a deployable boom and umbrella test and *Fuji-2*, an amateur radio satellite. The French ground station at Korou on the south American coast picked up the 780-kg *Momo* on its first pass and so, not long afterward, did Japanese stations at Katsura, Masuda and Okinawa. Within an hour the 5.2 m long solar panel had unfurled and the satellite began a 60-day period of checkout. MOS-1B carried a dish-shaped microwave scanning radiometer, an X-band antenna and a tube-shaped multispectrum electronic self-scanning radiometer. Its instruments were able to identify red tide (jellyfish infestations), the distribution of snow and ice and volcanic ash. Detailed colour maps of sea temperatures were compiled for the seas around the Japanese islands, blue for cold currents and red for warm sea. MOS-1B continued to operate until 1996 when it was closed down after a battery failure.

The 50-kg *Orizuru* was an unusual test. The concept was to test out, in miniature, the deployment of free-flying microgravity platforms from orbital stations. These would park some distance from the station, carry out experiments in zero gravity unperturbed by space station operations and then return. A special purpose of the test was to verify if umbrella devices could be used, combined with atmospheric drag, to manoeuvre back to an orbital station. In the course of a week, *Orizuru* deployed its boom 34 times and its umbrella 52 times in what was apparently a successful test.

JERS: INTRODUCTION OF SPACE-BORNE RADAR

The next step after MOS was a land observation Earth resources satellite. This was JERS, or the Japanese Earth Resources Satellite, the first radar-based Earth resources satellite. 96% of the satellite was produced in Japan. Weighing 1.4 tonnes, it was equipped with a 12 by 2.5 m Synthetic Aperture Radar (SAR) and multi-spectral imaging radiometers. Because of the technical complexities involved, the design period was unusually long— twelve years. It was launched successfully on 11 February 1992 from Tanegashima, the H-I heading due south toward a 568-km orbit, repeating its ground path every 44 days. Separation took place over Argentina 50 min after take-off and the satellite was duly renamed *Fuyo*.

Problems arose when the SAR failed to deploy because one of six pins holding the antenna in place stuck. However, it seems that extreme cold may have contributed to the problem, because when the satellite was pointed at the Sun several weeks later the pin suddenly popped open. Even when it did work, radar images were spoilt by stripes appearing on the pictures, a problem which was eventually overcome by programming them out during processing. When this was done, ground controllers were able to get razor-sharp images of Mount Fuji volcano and the Japanese islands. Later, its instruments were used for land surveys, studies of fisheries and agriculture, natural resources work and disaster prevention. *Fuyo* was able to track forest fires in Mongolia and the crustal movement of the Earth near Iwate volcano.

The main receiving centre for *Fuyo* information was the Earth Observation Centre at Hatoyama, the centre being responsible for processing, distribution and building an archive. The centre had three 11.5-m dishes to receive signals. *Fuyo* relayed stored data both to Hatoyama and the University of Alaska in Fairbanks and to a further ten Earth stations equipped to receive real-time data—Kumamoto in Japan, Bangkok, Showa Base in Antarctica, Fucino, Kiruna, Maspolamos, Tromso, Gatineau, Prince Albert and Beijing. By the late 1990s, the centre had built up considerable expertise through its handling of data from Japan's MOS 1 and 1B, JERS and ADEOS, Europe's ERS, the United States' *Landsat 5* and France's SPOT 2.

Fuyo was shut down in 1998, having far exceeded its design lifetime. By this stage, ground stations had received 90 000 photographs and 140 000 radar images. The batteries started to malfunction in October and then two of the three gyros switched themselves off. It was unable to acquire the Sun and, starved of electrical power, the electricity system failed. It was later formally switched off.

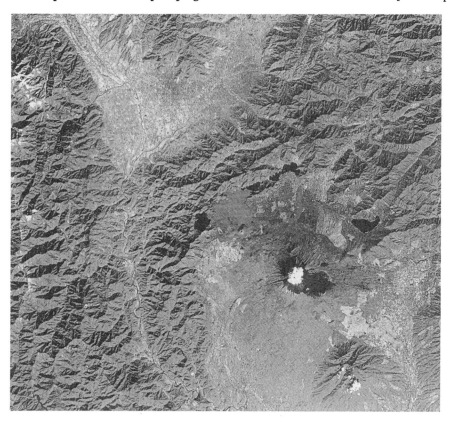

Snow-capped Mount Fuji from *Fuyo*

ALOS: DAY AND NIGHT, CLOUD-FREE

JERS will be replaced, after a gap, by ALOS, or Advanced Land Observation Satellite. This ¥36.35bn (€328m) project will be the largest Earth resources project ever devised in Japan, weighing 3850 kg and requiring 7 kW of electrical power, for which a single, long panel will be used. Set for launch in 2003, it will carry a stereo-mapping remote-sensing instrument using three telescopes called Prism with 2.5-m resolution, a 240-kg visible and infrared radiometer (AVNIR) with 10-m resolution and a 475-kg phased array synthetic aperture radar (PALSAR), able to provide day and night, cloud-free observations. ALOS will have a major role in mapping and disaster management, revisiting each site in the Japanese islands every two days. Its observation platform could also, experts later noted, be valuable in keeping track of military movements, not least North Korea's nascent.rocket programme. ALOS will orbit at 700 km every 100 min.

The data from ALOS will be used by the Geographical Survey Institute to draw up a new map of Japan and the Asian-Pacific region on a 25 000:1 scale, by the Japanese Environment Agency for environmental censuses and to promote sustainable development, and by the National Maritime Safety Agency for responses to natural disasters. It is

Ocean temperatures mapped from Earth orbit

intended that ALOS match and exceed the capabilities of the American *Landsat 7*, launched 1999, and the French SPOT 5, set for 2002. Data will be transmitted through American and Japanese relay satellites.

ADEOS/*MIDORI*: THE GREEN OBSERVER

ADEOS was the third Japanese remote sensing satellite, following *Momo* and *Fuyo*. The Advanced Earth Orbiting Observation Satellite was a 3560-kg gold-foil covered platform designed to note global changes in the Earth's satellite, atmosphere and oceans, in particular the ozone layers, the tropical rainforest, carbon dioxide at the poles and the greenhouse effect. ADEOS carried eight sensors, of which five were developed by Japan, two by NASA and one by the French space agency CNES. It was an expensive satellite, costing ¥112bn (€1bn) and measuring 8 m tall and 4 m wide. A small, 50-kg amateur radio satellite developed by Nippon Electric and the Japan Amateur Radio League was launched piggyback on the mission by the H-II on 17 August 1996.

The ADEOS spacecraft, renamed *Midori* ('green'), circled the Earth at nearly 800 km, encountered a routine range of teething troubles but was declared operational in November 1996. Data began to flood in to the NASDA Earth Observation Centre in Hatoyama. The first set of pictures were superb and presented in a variety of formats (true colour and false colour). They identified concentrations of chlorophyll, tropical storms off Japan (typhoons Violet and Tom), the El Nino current stretching across the Pacific and the southern ozone hole. The Improved Limb Atmospheric Spectrometer noted the rise and fall of ozone concentrations and another meter recorded the distribution of greenhouse gases. In early 1997, *Midori* tracked the spread of spilt oil from a tanker off Japan, the pictures being put up on the internet.

Important data back came from the AVNIR, or Advanced Visible and Near Infrared Radiometer. Although designed for soil, vegetation, pollution and energy studies, the instrument, with 8-m resolution, sent back outstandingly clear pictures of urban areas such as Hiroshima. Early pictures from the eight-band Ocean Colour and Temperature

Scanner provided images of plankton in the sea off the Japanese coast. ADEOS also carried a TOMS, one of the environmentally most important instruments in Earth observation. TOMS, or Total Ozone Mapping Spectrometer, was invented by NASA in the 1970s and when flown on *Nimbus 7* in 1978 it found the ozone hole over the Antarctic. A second TOMS was flown on a Russian *Meteor 3* satellite and ADEOS was its third assignment. Within a month, ADEOS had sent back new and worrying images of both the Antarctic and Arctic ozone holes. No wonder it was one of the heaviest payloads ever put into orbit by Japan.

Contact with the spacecraft suddenly ended on 30 June 1997. Solar power from its wide 30-m wings was quickly lost and the batteries drained in four hours. Space débris was blamed at first. The satellite was declared abandoned the following month. However, a subsequent investigation found a more mundane, more human and more likely cause: a weld had broken at the base of the solar panel.

A second, replacement, more specialized ADEOS 2 was ordered for 2001. The new ADEOS will carry an advanced microwave scanning radiometer with a wide swath (1600 km), an optical sensor working in 36 spectral bands able to look sideways as well as down and a limb atmospheric spectrometer which can search for pollution (aerosols and ozone) in the atmosphere. As well as these Japanese instruments, overseas countries are contributing a suite of equipment and NASA a sensor to measure sea winds and direction, while the French space agency CNES will contribute an instrument called Polder to measure how the atmosphere reflects solar radiation.

TRMM, *ORIHIME* AND *HIKOBOSHI*: THE NEW STARS OF VEGA

Early in the morning of 28 November 1997, the sixth H-II blasted off from Tanegashima in one of its most complex assignments. First, 14 min into its mission, it released its first payload, TRMM, and then, after 27 min, a second payload, ETS 7 or *Kiku-7*. It was also the first to be launched outside the limited 90-day launch window. Agreement to do so followed prolonged negotiations with the five fishermen's unions involved. The Science & Technology Agency already paid ¥400m (€3.57m) a year to the fishermen for harbour works and compensation for the 90-day window from Tanegashima and the extension of the window cost a further ¥300m ((€2.67m) compensation. TRMM entered orbit of about 350 km, circular at 35°.

The 3.6-tonne TRMM or Tropical Rainfall Measuring Mission was a joint NASA/NASDA mission, Japan supplying the all-important rain radar. It derived its tropical rainfall title from the key purpose of the mission, which was to provide the first comprehensive picture of tropical rainfall, which comprises three-quarters of the world's rain and atmospheric energy. TRMM had an unusual shape of boxes, trusses and rectangular containers. ¥52bn (€471m) TRMM carried a microwave imager to peer through clouds, a rain radar, high-resolution infrared scanner, lightning meter and a sensor to measure the Earth's energy. The rain radar could make a three-dimensional picture of rain over a swath width of 150 km with a resolution of 4 km. This was a giant instrument, the shape of a honeycomb, twice the height of a person and three times the length. It immediately proved its value in tracking typhoons and measuring their cloud height, information vital to an assessment of the danger they posed. The first images from TRMM were stunning—showing the band of tropical rain around the entire southern hemisphere, swirling cyclones (with heavy rainfall marked by false red colours) and three-dimensional cloud profiles showing where torrents of rain were falling down. Within 18 months, TRMM had improved global rainfall measurements by 25%. A long-term function of the mission was to estimate climate change. An early target was the El Nino current which had made such an adverse impact on the wester Americas. NASA paid €283m and NASDA €188m for TRMM.

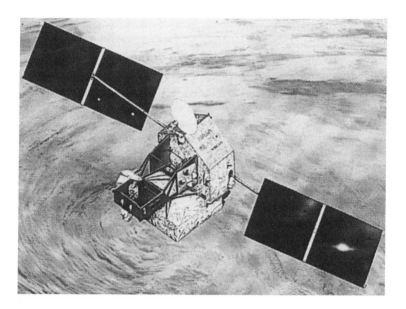

Tropical Rainfall Measuring Mission satellite

TRMM was launched as a companion to Engineering Test Satellite 7, ETS 7 or *Kiku-7*, costing ¥33bn (€245m). This was a radically different mission, one designed to explore Earth-orbiting rendezvous techniques that Japan would need to learn whenever it sent spaceplanes up to the International Space Station. Seven manoeuvres were planned.

ETS comprised two satellites—*Hikoboshi* (meaning Altair, the active satellite, weighing 2.5 tonnes) and *Orihime* (meaning Vega, the target, 400 kg), named after lovers in a Japanese fairy tale. *Hikoboshi* was box-shaped with twin solar arrays and 220 kg of fuel; *Orihime* was much smaller with one array. The programme called for *Hikoboshi* to follow a number of rendezvous and docking profiles using ground control, radar and the global positioning system. After rendezvous, *Hikoboshi* was to test its remote manipulator arm for typical operations that would be carried out on the Japanese module of the international space station.

The mission got off to a bad start, for it was hit by a gyroscope failure which prevented it from achieving proper orientation and stability was lost for a worrying six hours. Then attempts to connect ETS to the American tracking and data relay satellite system broke down. However, mission control in Tsukuba went ahead and masterminded the first rendezvous and docking tests which began on 7 July 1998—by coincidence the festival of Vega—when the two spacecraft separated to a distance of 2 m for 30 min. Using sensor and lasers to detect one another's position, the smaller satellite inched toward the mother craft, which then gripped it with three pincers 550 km above Earth.

On 6 August, *Hikoboshi* separated from *Orihime* to a distance of 2 m and redocked a few minutes later. Later that day, they separated to 525 m in preparation for more ambitious experiments. At this stage, attitude control on *Hikoboshi* failed and *Hikoboshi* retreated to a distance of 5 km while trouble-shooting took place. Two attempts to re-dock failed on the 8th and 9th, the satellites coming to within 110 m of one another. The satellites lost one another's position and seemed not to receive their instructions from Tsukuba. Again, on 13 August, the satellites came to within 145 m when attitude controls failed. After two further unsuccessful attempts when thrusters gave problems and attitude controls failed, *Hikoboshi* eventually pulled off a successful re-docking on 27 August, the re-docked combination then being in orbit of 542–444 km, 35.97°, 95.51 min. In achieving the re-docking, *Hikoboshi* used up most of its manoeuvring fuel. All this was televised as the Earth rolled below.

The docking was hailed by the Japanese and western press as the first ever unmanned docking [7] (which it was not, for the USSR had done this in 1967) and as an important step to testing out procedures for sending Japanese spacecraft to future space stations (a justifiable claim). Although the manoeuvres were full of difficulties, such as thruster anomalies and attitude failures, the purpose of the mission had been to trouble-shoot such problems. When they did occur, collision avoidance procedures were followed success-fully and ground controllers were able to park the chaser while new procedures were

Orihime and *Hikoboshi* separate—seen on TV

worked out and fresh software loaded up to the spacecraft. The thrusters and attitude control systems of NASDA's planned relay satellites for the International Space Station were radically overhauled as a result. The engineers responsible were showered with awards from the Society of Mechanical Engineers, the Society of Control Engineers and the Robotics Society of Japan.

This was far from the end of the *Hikoboshi/Orihime* experiment. The robot arm and hand on board, miniatures of those to be used on the International Space Station, were used to inject gas into a simulated tank, deploy and disassemble a truss structure, fasten bolts, connect electrical leads and capture floating objects. Communications were relayed through NASA's system of tracking and data relay satellites, far out in 24-hr orbit. In spring 1999, German Space Agency (DLR) engineers used virtual reality computers to make further tests of the arm and hand. NASDA astronaut Koichi Wakata came in at one stage to operate the remote arm. In September 1999, *Hikoboshi* and *Orihime* separated, though only to a 200-mm distance, and *Hikoboshi* was instructed to use its hand to recapture its companion without ground assistance. Despite the very small quantity of fuel remaining, and despite some propulsion anomalies, a further set of release-and-recapture tests were carried out on 27 October 1999. The two years of ETS-7 experiments were vital in laying the groundwork for Japan's participation in the International Space Station.

ETS 8: A GIANT, HOVERING INSECT

The next ETS, ETS 8, was scheduled for launch on H-IIA in 2002. ETS 8 will have large deployable reflectors, 19 by 17 m, twice the size of the Muses B experiment, each the size of a tennis court, making it look like a giant hovering insect. Deploying the antenna is of itself a difficult and delicate undertaking involving the development of carbon-fibre-reinforced struts and computer-controlled small electric motors to push the reflectors into shape, all tested out in zero-gravity flights in Airbus aircraft. The intention of the test and the reason for the large antenna, the largest such civilian array ever put in orbit, is to provide telecommunications for mobile phones from 24-hr orbit that have hitherto only been possible in low Earth orbit. ETS 8 will demonstrate high rates of transmission and broadcast CD-quality sound. Built by Toshiba and Mitsubishi, the ¥32bn (€292m) ETS 8

ETS/*Kiku* in summary		
ETS 1	9 Sep 1975	First test of N-I
ETS 2	23 Feb 1977	First satellite to geosynchronous orbit
ETS 4	11 Feb 1981	
ETS 3	3 Sep 1982	
ETS 5	27 Aug 1987	Test of communications
ETS 6	28 Aug 1994	Test of lasers, data relay, thrusters
ETS 7	28 Nov 1997	Rendezvous and docking tests
ETS 8	2002	Mobile communications from 24-hr orbit (scheduled)

is intended to ensure that Japan remains at the cutting edge of new telecommunications technologies, later paving the way for spacecraft with very high data rates. ETS 8 will weigh 6.5 tonnes at launch and the main spacecraft box is 3 tonnes, a third larger than any previous Japanese comsat.

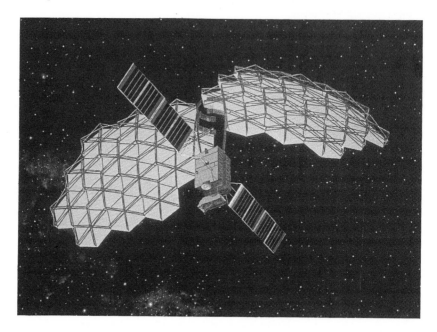

ETS 8—like a hovering insect

WINGED BIRD: COMETS/*KAKEHASHI*

A large, advanced communications engineering test satellite was launched on 21 February 1998. It was called COMETS—or COMmunications and broadcasting Engineering Test Satellite—and renamed *Kakehashi* ('bridge') in orbit. Constructed by NEC, COMETS was intended to use transponders to test communications with mobile phones and other satellites, demonstrate tracking and data relay systems, test high-definition television and the use of four xenon propulsion motors, each of 25 mN thrust. Like a winged bird, it had enormous solar panels 30 m long. The cost was ¥45.2bn (€403) and it was the most advanced communications technology satellite built in Japan.

The mission got off to a bad start when the H-II second stage shut down early, only 44 sec into a 192-sec burn, making it impossible for the satellite to reach its intended destination of geosynchronous orbit. The 3.9-tonne COMETS was left in a 247–1883-km orbit instead of the 36 000 km one planned, one of Japan's few launch failures.

Fortunately, of the 3.9-tonne weight of COMETS, 1.9 tonnes consisted of station-keeping fuel. In March, a 90-sec burn lifted the perigee from 250 km to 390 km, the first in a series of seven manoeuvres to raise the orbit to 500–17 700 km and thus enable about 60% of the original mission to be carried out. A NASDA investigation panel later found that a tiny hole had burned through the LE-5A's motor nozzle casing, igniting wires which caused the engine to shut down. COMETS did manage to relay signals for the ETS 7 automatic docking operations, but the mission as a whole could not be saved and was abandoned in mid-1999, two years ahead of schedule. Its orbit will last at least a thousand years.

Japanese launchers with date of first flight	
1970	*Lambda*
1971	*Mu-4S*
1974	*Mu-3C*
1975	N-I
1977	*Mu-3H*
1980	*Mus-3S*
1981	N-II
1985	*Mu-3SII*
1986	H-I
1994	H-II
1996	J-1
1997	*Mu-5*
2001	H-IIA (scheduled)

BEAMS ACROSS SPACE

The successor to COMETS is OICETS, or Optical Inter Orbit Communications Engineering Test Satellite, a small 550-kg communications engineering test satellite set for launch in 2002 planned jointly with the European Space Agency. On top of the spacecraft box is a laser, shaped like a camera with a zoom lens. Japanese engineers believe that if they can hold the communications beam steady, they can transmit large volumes of data up to 45 000 km through open space in narrow beams and s-band links. It will fire laser beams at the European *Artemis* satellite with a degree of accuracy of 0.0003°. OICETS will be the second and possibly last spacecraft to ride the J-I launcher and will enter a circular orbit of 550 km. Built by NEC, it will have a one-year operational life. OICETS is a box shape, has two solar panels and is 9.36 m long.

In a similar beams-across-space experiment, Japan joined the *Artemis* project, a ¥91bn (€810m), European Space Agency, 3.1-tonne, experimental technological satellite (Advanced Relay and Technology Mission Satellite). Designed to operate in geosynchronous orbit, *Artemis* was built with data relay systems to link to satellites in lower orbits, navigation transponders, lasers for inter-satellite communication and ion engines for station-keeping. When the project went over budget, the cash-short European Space Agency sought a free launch from Japan on the first H-IIA in exchange for NASDA being given 40% of the payload for data relays, pending the launch of Japan's own Data Relay Test Satellite west and east (DRTS W and DRTS E).

OICETS: beams across space

CONCLUSIONS: ADVANCES IN TECHNOLOGY, ENGINEERING AND APPLICATIONS

The 1990s saw significant Japanese advances in technology, engineering and applications. Three Earth resources satellites were launched. Engineering Test Satellites, carrying a number of cutting edge technologies in communications and robotics, were flown in Earth orbit. The launch of communications satellites continued apace, accelerated by the liberalization announced in 1989. Japan developed one of the world's most advanced rockets, the H-II and a sophisticated solid rocket, the J-1, both build indigenously, completing Japan's launcher family.

5

The Japanese in space

The various unmanned space ventures undertaken by Japan in the 1970 and 1980s, spectacular and adventurous though they were, lacked a human presence. Public interest in spaceflight is always higher when men and women are directly involved, with human drama and astronauts putting their lives at risk in often dangerous adventures. In the 1990s, Japanese astronauts began to fly in space on the shuttle and Japan became an important contributor to the world's most ambitious scientific programme—the International Space Station.

JAPAN'S FIRST ASTRONAUT

Japan was a longstanding economic and scientific partner of the United States, reflecting the close ties established in the postwar years. Ultimately, Japan always wanted to have astronauts in orbit performing experiments of interest to Japanese science, although it recognized that manned Japanese space shots were a distant prospect. However, the American space shuttle offered the prospects of a Japanese person reaching Earth orbit in the late 1980s.

The space shuttle had a large payload bay. This was intended to carry satellite payloads into orbit, but could also house a small laboratory where scientists could work for a week at a time, then the limit of a shuttle mission. This was called *Spacelab*. The United States were financially very constrained by the high costs of building the shuttle, so they invited their international partner nations to share in the cost and building of a shuttle laboratory, the *quid pro quo* being that their astronauts could fly on board the shuttle.

The *Spacelab* was a mini-space-station module built in Europe and designed to fit snugly into the payload bay of the shuttle. It was actually a combination of modules that could be configured in a number of different ways, depending on the mission required, the standard being a pressurized module with ready-to-fit equipment racks. Connected by a tunnel to the main cabin of the space shuttle, the idea was that astronauts would float down the tunnel to conduct experiments in a shirtsleeve environment. *Spacelab* was equipped with experimental racks, containers and platforms, and could be adapted according to the type of mission required. Between 1983 and 1998, NASA was to fly *Spacelab* 22 times, kitted out for different countries (Europe, Germany, Japan) or for different sets of scientific experiments (e.g. microgravity, materials science experiments). Japan was to be involved in four spacelab missions. One was a dedicated Japanese space mission, *Spacelab J* and Japanese experiments flew on three other *Spacelabs*, including the first one.

Japanese equipment flew on the first *Spacelab* mission (though not a Japanese astronaut). In advance, NASDA flew a number of materials processing experiments on sounding rockets to pave the way for melting and crystal growth studies to be carried out on *Spacelab*. The first sounding rocket, called the TT-500A, was fired on 14 September 1980 from Tanegashima. The TT-500A climbed 320 km into the Pacific sky, providing 7 min microgravity as the payload described a giant arc 500 km downrange.

Tracked by the Ogaswara station, parachutes deployed at 6 km, with splashdown 16 min after launch. However, the second sounding rocket launch on 15 January 1981, carrying a metallic compounds experiment called Ni-TiC whisker, was lost when the beacons failed on splashdown.

In December 1982, the first main *Spacelab* crew and their ground supports—Byron Lichtenberg, Michael Lampton, Ulf Merbold, Wubbo Ockels, Owen Garriott and Robert Parker—travelled to Japan to operate the equipment in the NASDA vacuum chamber outside Tokyo. *Spacelab* flew the following year carrying a Japanese–American equipment, designed by ISAS, NASA and the Texas South West Research Institute. The Space Experiments with Particle Accelerators were designed to help scientists better understand the relationship between the magnetosphere and the upper atmosphere by injecting charged particles into plasma.

Japan then negotiated a deal with the United States for a *Spacelab* mission to be devoted entirely to Japan. Japan would kit out the lab and set up experiments which would be operated by Japanese astronauts. The mission was called *Spacelab J* (J for Japan, comparable to *Spacelab D*, dedicated to Germany (D for Deutschland)). The mission was first set for 1988 on the space shuttle *Challenger* with the title of mission STS-81G. This required the selection of Japan's first group of astronauts and a call for candidates duly went out. Three Japanese astronauts were selected from 533 applicants: Dr M Mamoru Mohri, Dr Chiaki Mukai (both NASDA) and Dr Takao Doi.

Mamouri Mohri was a chemistry scientist, born in 1948, who graduated from Hokkaido University and subsequently studied in Australia. Chiaki Mukai was born 6 May 1952, went to a girls' high school and graduated from Keio University School of Medicine in 1977, acquiring a doctor's licence with a specialization in cardiovascular surgery. Takao Doi, born 18 September 1954 in Minamitama-gun, Tokyo, was a engineer with primary, masters and doctoral degrees from the University of Tokyo. 44-year-old Mohri was selected for *Spacelab J*, which was renamed *Fuwatto*, a Japanese word for 'floating free' or 'weightless'.

For the first mission, 34 experiments were selected—22 in the area of materials processing and 12 in life sciences (all but three were Japanese). The former would build on the work carried out on the TT-500A sounding rockets fired in the early 1980s.

The *Spacelab J* mission was set for January 1988, but delayed indefinitely when the space shuttle *Challenger* exploded over Cape Canaveral on 28 January 1986. The Soviet Union offered to fly the intended experiments on board its new *Mir* space station launched a month later, but the Japanese decided to keep to their arrangements with the United

States, even though it meant a long delay. The shuttle programme got on track again in late 1988 and the first parts of *Spacelab J* arrived at Kennedy Space Centre just over two years later.

INSTEAD, A MISSION TO *MIR*

In the event, Japan's first astronaut flew on a Russian, not an American spaceship. In 1989, the Tokyo Broadcasting System decided to celebrate its 40th year of broadcasting by sending one of its own journalists into space. TBS was the largest private broadcaster in Japan, with 26 stations, 1600 staff and 8000 affiliate staff, with 16 overseas bureaux. The station had a longstanding interest in spaceflight. TBS had been the second western company permitted to film at Baikonour cosmodrome and had covered the flight of the Soviet space shuttle, *Buran*.

TBS approached the Soviet space agency, Glavcosmos, and struck a deal whereby TBS would pay ¥1.26bn (€11.3m) to fly a journalist to the *Mir* space station for a week (much cheaper than what NASDA paid NASA to fly Japanese astronauts on the shuttle, ¥3bn each). In the event, TBS sub-sponsored its own costs, inviting advertising from Sony (whose letters were displayed prominently on the rocket at take off), Minolta, American Express Japan and various healthcare, chemical and insurance companies.

In a competition to find an appropriate reporter, 145 men and 18 women applied, ranging in age from 23 to 55. The applicants included newscasters, announcers, correspondents, TV directors, field staff and even members of the sales and accounts departments.

The first round of medical tests reduced the field to 21 people. Four Soviet doctors then arrived to assist in the next stage of selection, which reduced the group to seven. They were sent to centrifuge training. All failed: the Soviet doctors were adamant that they would not lower their standards just for this mission. One applicant had a stomach condition that produced gastric juices when under stress; another had blood vessels that were too fine. A particular problem for the Japanese was that 80% of the population have a deviated nasal septum (a bend in the inside channel of the top of their nose); whilst the Japanese did not regard this as a problem, conventional Caucasian science was that this imperfection invited infection.

So a second round of recruitment was organized, this time attracting 55 male and nine female applicants. They were exercised on bikes, made to lie down with their heads at a lower angle than their body, decompressed and whirled around in centrifuges. Following the tests, the Soviet doctors again expressed their unhappiness.

The Japanese doctors then went back to the first recruitment round and selected a different group of seven finalists: Moscow correspondent Toshio Koike, Toyohiro Akiyama, Ryoko Kikuchi, Atsuyoshi Murakama; and from affiliated TBS companies Nobuhiro Yamamouri, Kouichi Okada and Naoki Goto. Once again, this group was sent to the doctors and the centrifuges. This time, Soviet doctors found two who met their standards: 48-year-old editor Toyohiro Akiyama and 26-year-old camerawoman Ryoko Kikuchi. Their selection was duly announced in September 1989. Koike, Yamamouri and Goto were also allowed to train, provided that they had their tonsils taken out. They all went to Moscow on 1 October 1990. Murakami and Okada were dismissed.

Akiyama, born in 1942, was a sociology graduate who had previously been Washington bureau chief and a reporter in Vietnam, and had worked in the Japanese service of the BBC in London. Now he was a senior man within the company and editor-in-chief of the foreign news division. He had been commentator on a number of shuttle missions, including the *Challenger* disaster. Getting to the final selection was a major challenge for him, for he had to cut out two typical journalistic habits of drinking and smoking, in his case four packets of cigarettes a day. Kikuchi was much the fitter of the two, her recreational choices being mountain climbing, cycling, skiing, basketball, swimming and kabuki theatre. Born 1964 in Zama, Kanagawa, she had studied Chinese at the Tokyo University for Foreign Studies. She had less overseas journalistic experience, limited to China and the Seoul Olympics in Korea. The mission of the journalist was to make two 10-min television broadcasts each day, several 20-min radio broadcasts and contact with amateur radio hams. In the course of the broadcast, film would be taken of the behaviour of six Japanese tree frogs in orbit. Six television cameras were delivered in advance by the unmanned *Progress M-5* cargo ship.

The timing of the mission had never been entirely clear, 1992 being suggested as the likely date. At the time of the selection of Akiyama and Kikuchi, the *Spacelab J* mission had been set for June 1991. The Soviet Union brought the journalist mission forward to December 1990, in what was considered by observers as a deliberate move to ensure that the first Japanese in space rode a Soviet, rather than an American, rocket. The setting of the date did not—publicly at least—upset NASDA, who congratulated TBS on its achievement, but it enraged Soviet journalists who protested that a foreign journalist would fly in space before one of their own.

On 2 November 1990, the final team was chosen, Akiyama emerging as the winner. Even had the choice been otherwise, Kikuchi would have been robbed of her flight in any case, for she fell victim to an appendix three weeks later, just days before the scheduled mission.

Toyohiro Akiyama was duly launched on 2 December 1990 on *Soyuz TM-11* with Viktor Afanasayev and Musa Manarov. A hundred Japanese media technicians attended the launch, at a cost estimated at equivalent of that of the mission itself. Cameras were set up close to the rocket and its flame trench to give spectacular views of ignition. Emblazoned in Soviet and Japanese flags, *Soyuz TM-11* soared into a clear sky with live coverage on Japanese television. Akiyama carried with him a small Japanese mascot—a doll dressed in a bright kimono. Going on air as he entered orbit, the journalist radioed to Earth, echoing the famous first words from orbit of Yuri Gagarin: 'This is Akiyama! The Earth is blue!'. Akiyama was the 237th person in space.

Things did not go quite as planned. Soon after getting into orbit, Akiyama had a bout of space sickness, a problem which afflicts half of all space travellers. Akiyama was ridiculed in the, probably jealous, Western press as a chain-smoking, whiskey-swilling idiot. In fact, Akiyama shot some of the best film ever taken of life on board *Mir*; he conveyed a sense of what many ordinary people must feel in orbit; and his medical parameters were comparable to his professional cosmonaut companions. TBS attracted record viewing figures during the week-long mission. He came back with stunning colour film which made a top-class video.

In the course of his eight-day mission, Akiyama made 30 hours of live broadcasts from the space station, much of it timed for peak viewing time. He filmed activities inside the station and pointed his cameras outside the station to pick out landmarks below, such as Mount Fuji. He noted how small Japan was, compared to the large land masses of Africa, the Americas and Siberia. He returned to Earth on the *Soyuz TM-10* spacecraft with Soviet cosmonauts Gennadiy Manakov and Gennadiy Strekhalov who were coming back after four months in orbit. The retro engine fired at 5.13 a.m. on 10 December and the cabin reached the ground at 6.08 a.m. near Arkalyk, Kazakhstan. Emerging from the capsule, Akiyama announced that he was hungry and in need of a beer and some cigarettes. He had been in orbit for 7 days 21 hours and returned to his broadcasting career in Tokyo two weeks later.

This was not the end of the relationship between Japan and *Mir*. In 1996, under an agreement between NASDA and the Russian Space Agency, RKA, two pieces of equipment were later installed on the *Mir* space complex—one outside to test the effects of space radiation on cell cultures (radiation monitoring) and the other on the inside to examine how tiny organisms reproduced in weightlessness (microflora). Japan paid ¥1bn (€943 000) for the experiments.

FUWATTO'S SUCCESS

The way was now clear for the *Spacelab J* mission, or *Fuwatto*. The Japanese had now waited a long time for this mission and, for America's partners, the programme had proved an expensive and frustrating one. The numbers of *Spacelab* missions were restricted because the shuttle never achieved the frequent launch rates originally projected. The number of visiting, non-American astronauts was limited to one or two per mission. Even the studiously reticent NASDA hinted its disapproval of the unequal arrangement when it noted that *Spacelab J*, paid for by Japan, carrying entirely Japanese equipment, was operated by four Americans and 'only one Japanese payload specialist will be allowed to board the shuttle' [8]. The Americans later responded to their partners' unease by making many of the later American spacelabs quite international by nature and flying a number of visiting scientists.

44-year-old Mamouri Mohri was selected for the first *Spacelab J* mission. He completed his training for the mission in summer 1992. The mission duly took place on 12 September 1992 on the second flight of the space shuttle *Endeavour*, the mission designation being STS-47. Commanded by Hoot Gibson, Mamouri's companions were pilot Curt Brown, mission specialists Jan Davis, Jay Apt and Mae Jemison (the first black woman in space) and payload specialist Mark Lee. *Endeavour* climbed into a 280-km orbit of 57°, one which passed over Tokyo. They returned to Cape Canaveral on 20 September after their seven-day mission had been extended by one day to permit additional experiments. Mission commander Hoot Gibson later described it as a trouble-free mission. *Spacelab J* carried 30 kg of souvenirs—a flag of the rising sun, commemorative rubber stamps, the NASDA flag, and '*Fuwatto* '92' emblems.

The life sciences experiments concentrated on investigations of human health in space, but a quantity of other animals was also carried: two carp fish, 36 chicken embryos, 7600 fruit flies, 1800 hornets, fungi, plant seeds, frogs and frogs' eggs. Other experiments

Japan's shuttle astronauts—Mamouri Mohri, Chiaki Mukai, Koichi Wakata, Takao Doi

using a continuous heating furnace concerned gas evaporation in low-gravity, low-level acceleration and protein crystal growth. Jemison relayed video pictures of tadpoles hatching under weightlessness. Mohri gave a microgravity lesson to Japanese school children. For Gibson and Mohri, there was a TV relay to the Japanese prime minister and the head of NASDA. In order to get maximum benefit from the spacelab, two 12-hr shifts were worked with two astronaut teams, the red and the blue teams.

Mohri was in space for 7 days 22 hours and was the 282nd person to go into orbit. As soon as *Endeavour* landed, the life sciences experiments were removed—such as fruit flies, plant seeds and eggs (some hatched out later). There was a panic when the spacelab's electricity failed after touchdown and the carp were left without fresh oxygen for 27 min. But they survived. All the spacelab racks were shipped back to Japan a month later and post-flight analysis of the mission began.

The success of the *Fuwatto* mission was widely hailed in Japan. Although no other dedicated *Spacelab J* mission was in prospect, there was the possibility of more seats being available for NASA on future *Spacelab* and shuttle missions and later, on the international space station. Accordingly, a second call for astronauts was issued. In the second round of selection, held in 1991, 372 people applied to be astronauts. The requirements of candidates was that they should be Japanese, be less than 35 years old, speak English, be a graduate in natural science and have three years' research experience, be between 149 cm and 193 cm in height and give a ten-year commitment. Applicants were advised that they could expect to fly in four or five years on a shuttle mission before an operational assignment to the Japanese module. Of the 372 applicants, 331 were men and 41 were women. Japan's fourth astronaut was selected—Koichi Wakata, a 28-year-old Japanese Airlines engineer with a degree in applied mechanics.

In a supplementary recruitment, Soichi Noguchi was selected in a once-off round of astronaut election in May 1996, the third round. Born in 1965, Soichi Noguchi was an engineer, with degrees from Tokyo University and experience working for Ishakawajima-Harima Heavy Industries. He was chosen from a selection of 572 applicants and was sent straight after his selection to NASA Johnson Space Centre in Texas to train as a mission specialist. Later, in 1998, he became the first Japanese astronaut to train in the Yuri Gagarin Space Centre in Star Town, Moscow.

INTERNATIONAL MICROGRAVITY LABORATORY 1: NEWTS, FISH, CELLS

The International Microgravity Laboratory (IML-1) spacelab flew on the shuttle *Discovery* in January 1992. Japan provided two of the 42 experiments—one for crystal growth for the organic superconductor and the monitoring, and the other for the investigation of biological effects of cosmic radiation in a spaceflight environment.

Japan did get a seat on the second international microgravity laboratory spacelab, also called shuttle mission STS-65 (Space Transportation System 65). The honour went to Chiaki Mukai, who flew for 14 days 17 hours on the first of the shuttles to fly, *Columbia*. This mission, launched on 8 July 1994, flew 80 experiments, of which 12 came from Japan. The laboratory carried newts, fish and cell cultures; on the materials side, a large isothermal furnace and free-flow electrophoresis unit. In the course of the two-week mission, the newts laid eggs and the medaka freshwater fish spawned in orbit.

Chiaki Mukai later flew again on STS-95 on *Discovery* on 29 October 1998. This was popularly known as the 'John Glenn flight' because of the presence on board of America's first astronaut to enter orbit, John Glenn, in 1962. This time, Glenn flew partly as a test of how older people could survive a spaceflight, partly (though this was not admitted) to regenerate public interest in space flight. Chiaki Mukai's role was to operate five Japanese payloads on the flight in the *Spacelab* module—in effect, a mini-spacelab. These involved a study of neural signals in toadfish so as to better understand space sickness, the role of gravity in the growth of cucumbers, the use of melatonin to enable astronauts to sleep better and the test of blood samples.

SPACE FLIER UNIT

The concept of a free flier dates to the 1980s and approval for a Japanese free flier was given in 1986. In the same year, Japan set up the Institute for Unmanned Space Experiments with Free Fliers, a joint project between the Ministry of International Trade and Industry and 13 leading industrial and technological companies. The concept of a free flier is to launch a space platform on one spaceship which, after a period in orbit, is retrieved many years later by another. Ideally, a free flier is reusable and may be kitted out for a range of different missions.

First Japanese astronaut to fly twice—Chiaki Mukai

The four-tonne Space Flier Unit (SFU) was launched by H-II from Tanegashima on 18 March 1995. It was planned for later recovery by the American space shuttle. The flier took the form of a platform 4.46 m wide and 2.8 m high, with six payload bays on top. Electrical power was provided by two 24-m solar panels. During its period in orbit, thrusters maintained attitude control and adjusted its orbit, while an S-band antenna relayed data down to Earth control. This was not the first time that a space platform had been put into orbit for subsequent shuttle recovery, for this had been done with the Long Duration Exposure Facility in the 1980s. Japan's intention was to relaunch the SFU for further microgravity experiments five times. Also launched with the SFU was *Himawari 5* metsat.

The ¥60bn (€535m) SFU carried experiments in astronomical observations, the laying of eggs by water lizards and the growing of diamond crystals in weightlessness. The unit carried four experiments to test the possibility of transmitting solar energy from space to power electricity systems on Earth. Following ideas proposed by Peter Glaser for solar power systems, a group of enthusiasts began to explore how such ideas could best be developed in Japan, an important point given Japan's dependence on foreign raw energy sources. These were tested out first on sounding rockets and then the SFU. During its mission in Earth orbit, the space flier was able to survey 7% of the sky using an infrared telescope, providing useful information on zodiacal light, interstellar dust and background cosmic radiation.

The Space Flier Unit was retrieved by *Endeavour* (STS-72), launched 11 January 1996. The shuttle entered orbit almost 20 000 km behind the flier but it closed in rapidly. Two days later mission commander Brian Duffy steered the shuttle in for the final approach while mission specialist Leroy Chiao used a laser to call out precise distance data. Astronaut Koichi Wakata was on board (the 341st person to go into orbit). He was the third Japanese to ride the space shuttle but the first mission specialist. During the final approach, he waited in the back of the flight deck, ready with the remote manipulator arm to capture the free flier.

Then problems began. In advance of retrieval, ground controllers at Sagamihara Operations Centre near Tokyo commanded the 10-m solar arrays on the flier to retract. After many further commands, they still refused. Accordingly, a drastic backup plan had to put in place: explosive bolts were fired to blast the panels away as the shuttle and flier, in formation, passed the west coast of Africa. An orbit later, the shuttle closed in for the final manoeuvre. Wakata extended the arm to the now panel-less flier, grappled it with the remote arm lever and pulled it into the payload bay where it was safely berthed for return to Earth.

Wakata put his experience with the arm to further use the following day to release a NASA free flier. Unlike the Japanese one, this free flier called OAST or Office of Aeronautics and Space Technology, was for a short-term deployment and was recovered by Wakata after two days. It had been intended to fly the SFU up to five times, but this seems to be the only mission it is every likely to make.

PREPARING FOR THE INTERNATIONAL SPACE STATION

Japan's next astronaut was 43-year-old Dr Takao Doi, who flew on mission STS-87 on 19 November 1997 on the space shuttle *Columbia*. He was the fifth Japanese in space, the fourth on the shuttle, the first Japanese to walk in space and the 367th person to enter orbit. His companions were Kevin Kregel, commander, pilot Steven Lindsey, mission specialists Winston Scott and Kalpana Chawla and payload specialist Leonid Kadenyuk of the Ukraine. STS-87 was a dedicated microgravity mission, but the highlight of the mission was the spacewalk by a Japanese astronaut to test assembly techniques for the International Space Station. The spacewalk was modified because of a problem experienced by *Columbia* in retrieving the free-flying *Spartan 201* satellite and this had to be dealt with first. Winston Scott and Takao Doi donned spacesuits on 24 November, the sixth day of the mission, clambered down the payload bay and stationed themselves to grab the *Spartan* as the shuttle manoeuvred close. Television cameras showed them waiting for the right moment as the blue Earth rolling by filled the top half of the screen. They were able, with the help of the shuttle's remote manipulator system, to lock the *Spartan* down. This key task accomplished, they proceeded to test out a crane of the type which will be used to move equipment—some of it awkward and heavy—around the

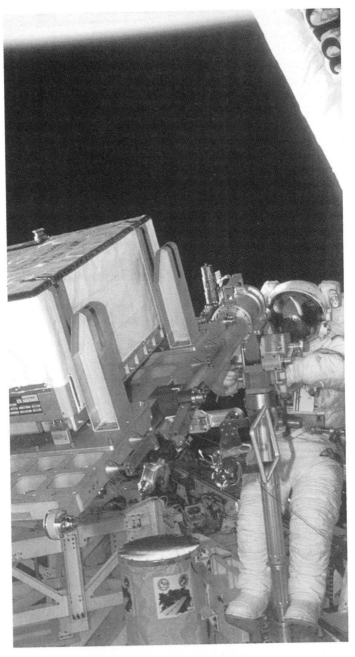

First Japanese space walk—Takao Doi

International Space Station. The astronauts returned to the cabin after 7 hours 43 min, one of the longest spacewalks done from the shuttle.

In order to gain more space station assembly experience, a second spacewalk was added to the mission on the 3 December. Scott and Doi went out for 5 h to again practice moving objects around the payload bay. In the course of the spacewalk, one of the astronauts in the cabin used a joystick to manoeuvre a beach-ball-size subsatellite called Aercam Sprint with two television eyes, designed to hover around and inspect difficult-to-reach parts of a space station. The shuttle returned to Earth after over 15 days with a haul of microgravity research.

Mamouri Mohri flew a second mission on the radar-carrying ST-99, while Koichi Wakata was selected for mission STS-92, the third mission to the international space station, scheduled for 2000.

JAPAN AND THE INTERNATIONAL SPACE STATION

In 1985, Japan was invited to take part in the project for the American Space Station *Freedom* announced by President Ronald Reagan the previous year. Japan was asked by NASA to contributed a pressurized module providing a shirtsleeve environment for a crew; an exposed facility; a scientific equipment airlock; a local manipulator arm; and a 3.1-tonne experi-mental logistics module. The module would carry equipment for 30 experiments in scientific, microgravity and materials processing research. The components would be launched by the American shuttle in its payload bay. The cost of participation was then estimated at ¥310bn (€2.76bn).

Japan quickly began preliminary design of its Japanese Experiment Module, the 13.4-tonne JEM, in 1987, later to be called *Kibo*.

On 14 March, 1989 a memorandum of understanding (MOU) was signed between Japan and the United States through their respective space agencies, for the 'design, development, operation and utilization' of the space station *Freedom*. Similar MOUs were signed at around the same time between the United States on the one hand and Europe and Canada on the other. By 1995, construction of an engineering module had been completed and in the following year the JEM passed a week-long critical design review. The review was attended by astronaut Koichi Wakata, who had just returned from orbit. The reviewers closed 344 review item discrepancies and sent 14 onward for further evaluation. The review process gave the go-ahead for manufacturing.

In Tsukuba Space Centre, a Space Station Operations Facility (SSOF) was completed in 1996 to support the work of the JEM. While overall control of the space station will rest with mission control in Houston, Texas, each country was responsible for the control of its own modules. The SSOF comprised a 24-hr operation control room, astronaut training centre, space experiment laboratory, space station test facility and zero-gravity test centre. The astronaut training centre comprised a 10-m deep swimming tank, an isolation chamber and a high-altitude chamber.

Although Japan carried out its preparatory work with exemplary efficiency, the same could not be said for the project leader. *Freedom* went through an endless series of redesigns between 1984 and 1992, eventually exhausting all the money allocated for its construction, although not a single component was ever built. The many redesigns caused nothing but frustration to the United States' international partners, whose designs re-mained sound despite the many American changes.

Newly elected President Clinton called a halt to the chaos in 1993 and ordered a final, cost-conscious redesign. This too came in over budget. At around the same time, the Russian *Mir-2* space station was becoming delayed by Russia's accelerating financial crisis and the only way to save both was by an effective merger of the two programmes, a decision reached remarkably quickly by both countries. A new Japanese–American international agreement was initialled to take account of the new arrangements and this was signed by Japan in November 1998, just before the launch of the first module of the station, the Russian *Zarya*. What had been the American space station *Freedom* was renamed, more neutrally, the International Space Station (ISS)

Despite the many ups and downs of its design and development, the building of the *Kibo* module remained steadily on course throughout the period from approval in 1989 to arrival at the Tsukuba space centre in 1997. For Japan, the ISS in general and its own module in particular offered the opportunity for permanent participation in manned space flight (albeit as a partner rather than a leader) and a platform where research could be carried out into manufacturing technologies in weightlessness and vacuum.

Japan also proceeded methodologically with the preparation of scientific experiments for the space station. The first call for scientific experiments was made in 1992 and led to the selection of the first 49 Japanese experiments. Once the station is up and running, proposals will be invited annually. The invitation will stress the importance of experiments being done speedily and taking not more than half a day's supervision. In 1997, the

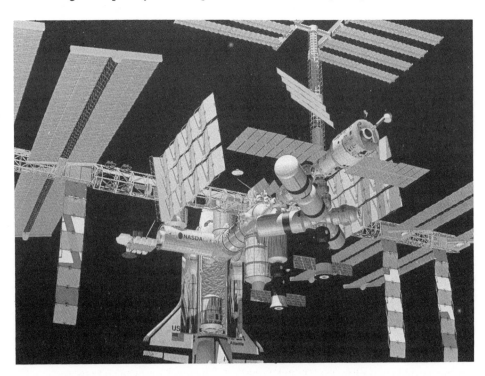

The International Space Station: the NASDA module is slightly to the left of the centre

first payloads were selected for the Exposed Facility—a pulsed laser beam to detect space débris smaller than 10 cm, an X-ray detector, an instrument to study trace gases in the atmosphere and a high-capacity space communications system. The experiments to be run on *Kibo* will be operated by the Space Environment Utilization Research Centre in Tsukuba.

In advance of the launch of JEM, sounding rockets were used to test some of the experiments to be done on board the space station, the first in September 1991. On 20 August 1992, the TR-1A2 sounding rocket was fired to 227 km to carry out preliminary microgravity experiments of the type that would later be flow on the JEM. On 25 September 1996, NASDA launched the third TR-1A test rocket from Tanegashima to test six microgravity experiments. The sounding rocket's engine burned for 64 sec, reached an altitude of 262 km, the payload splashing down in the Pacific ocean, 14 min after lift-off, 320 km downrange. The launching offered 361 sec of microgravity, permitting the study of the growth of colloidal crystals, nucleate boiling, the combustion of air–fuel mixtures, the melting of germanium and welding. A fourth JEM precursor sounding rocket was flown 25 August 1995.

CONSTRUCTION

Although the *Kibo* module is the main Japanese contribution to the station, first up there will be the logistics module, a 3.9 m tall, 5-tonne cylinder to be placed atop *Kibo*. Originally this was intended as a full-size research module, but this had to be downsized when the station took a higher orbit as a result of Russian participation. The logistics module became a storage facility with eight equipment racks instead.

Kibo, the pressurized module, or PM, is the centre of Japanese participation in the ISS. In appearances not unlike *Spacelab J*, the 15.3-tonne module is 11.2 m long, carries 23 experiment and storage racks and can be used by up to four astronauts. Two Japanese astronauts will work in this module at any given time. Running *Kibo* will require between 5 and 25 kW of power. Main contractor is Mitsubishi Heavy Industries. Launch is set for 2002. Among the experiments anticipated in the pressurized module will be the development of semiconductors and alloys, the creation of new materials and medicines, the performance of liquids in weightlessness and life sciences.

One of the first objects to be designed was the JEM Remote Manipulator System (JEMRMS), a robot arm that would be used to move large and small payloads around on the outside of the facility. The JEMRMS comprised a 10 m long large arm able to handle 7 tonnes and a 1.5 m fine arm planned for lighter objects up to 700 kg. The main arm will launch with *Kibo* and the small arm with the exposed facility. The fine arm was tested out successfully on the shuttle mission STS-85 in August 1997: astronauts Jan Davis and Steve Robinson put the system through its paces over four days of tests, learning to move objects as it would do on the station.

The exposed facility or EF weighs 3.8 tonnes and is attached below the airlock of the pressurized module. Using the remote manipulator arm from the inside of the module, Japanese astronauts will be able to conduct experiments outside, whilst still in a shirt-sleeve environment in the comfort of the module. The platform, which measures 5 by 5.2 by 4 m, can take payloads of all different shapes and sizes and is suited for up to four

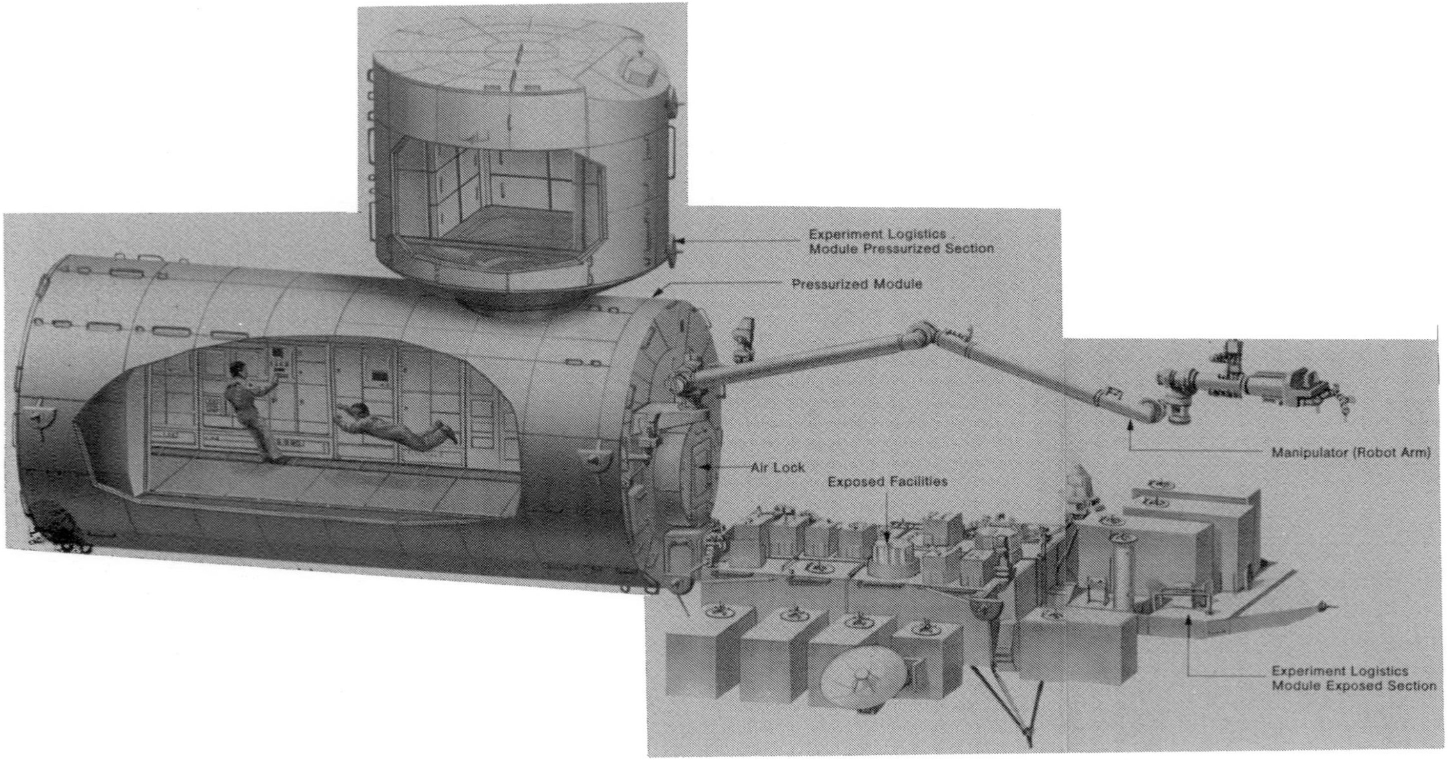

The *Kibo* module, arm, exposed facility and logistics module

Experiment Logistics Module Pressurized Section

Pressurized Module

Air Lock

Exposed Facilities

Manipulator (Robot Arm)

Experiment Logistics Module Exposed Section

plug-in modules. Launch is set for 2002. The main experiments to be conducted on the outdoor porch will be astronomical and materials processing.

Making JEM operational will involve a number of follow-up missions:

- resupply by unmanned resupply ship, the HOPE transfer vehicle (HTV)
- new astronauts, to be boarded by the American shuttle
- two relay satellites to provide round-the-clock ground control contact with the station
- later, the introduction of the HOPE shuttle.

In 1999, Japan began studies with the United States on a joint 10-tonne centrifuge module to add to the station and for an eventual upgrade to the JEM for 2010 to 2030 (the JEM was a 1988 design and risked becoming outdated earlier than expected).

KEEPING IN CONTACT: DATA RELAY

The DRTS is an important part of the operation of the Japanese part of the International Space Station. Early manned spaceflight suffered from problems of limited communications—astronauts could communicate with the ground only when they passed, generally quite quickly, over ground or sea tracking facilities (though, ironically, this was not a problem with the Apollo Moon landings which, although they were far away, were always in line of sight with the ground). Early in the shuttle programme, NASA introduced the tracking and data relay system, whereby shuttles communicated outward to 24-hr communications satellites, which relayed voice and data back to mission control in Houston. With three such satellites, it was possible for the shuttle to communicate with the ground throughout each flight. The Soviet Union introduced a similar system for *Mir* called *Luch*, although owing to financial shortages it operated only sporadically. For ISS, constant communications with the ground will be essential and data relay satellites will be necessary to make this possible.

Japan made its first studies of a relay system in 1988 at the Tsukuba Space Centre. For *Kibo*, Japan will operate a system called DRTS, Data Relay and Tracking Satellites called E and W respectively, because they will be positioned at 90° east and 160° west respectively. This will provide almost global coverage, enabling *Kibo* to send down 50 megabytes of data per second and take 3 megabytes per second up. These 2-tonne spacecraft will be based on the experience gained on the ETS 6 mission and first entered the budget in 1990. Each will have a large 5-m dish antenna.

MODULES BUILT, ASTRONAUTS RECRUITED

In 1998, a mockup of the ISS, including the Japanese JEM, had been assembled in the Mockup and Integration Laboratory in Houston, Texas. Visitors could walk into the spacious module, admire the wide range of experimental equipment and storage racks installed on the walls and note the large airlock designed to facilitate access to the exposed facility. They could examine the robotic arm on the outside of the module and the drum-shaped logistics module on top. The module was painted with a red image of the Japanese flag, accompanied by the letters 'NASDA' in English and Japanese.

As part of the human preparation for the ISS, Japan agreed to participate in an eight-month experiment at the Institute for Biological and Medical Problems in Moscow. Japan sent one of its trainee astronauts, Soichi Noguchi, to join four Russians astronauts from the United States, France, Germany and Austria in a simulated flight to the ISS to see how an international crew would live in the confined space of the future laboratory.

Nine Japanese companies have a major involvement with the International Space Station, led by Mitsubishi (Heavy Industries and Electric) which is committing ¥310bn (€2.7bn). The others are Toshiba, Ishakawajima, Nissan, NEC, Kawasaki, Hitachi and NTT.

Japan's astronauts

Flown

2 Dec 1990	Toyohiro Akiyama	*Soyuz TM-11*
12 Sep 1992	Mamoru Mohri	STS-47/*Spacelab J*/*Endeavour*
8 July 1994	Chiaki Mukai	STS-65/*Columbia*/IML-2
11 Jan 1996	Koichi Wakata	STS-72/*Endeavour*/Space Flier
19 Nov 1997	Takao Doi	STS-87/*Columbia*, 1st EVA
29 Oct 1998	Chiaki Mukai (2)	STS-95/*Discovery*
11 Feb 2000	Mamoru Mohri (2)	STS-99/*Endeavour*

Selected but not yet flown

Ryoko Kikuchi	*Soyuz TM-11*	Trained in Russia, returned to broadcasting
Noguchi Soichi	ISS	NASDA, selected 1996
Naoko Sumino	ISS	NASDA, selected 1999
Akihiko Hoshide	ISS	NASDA, selected 1999
Satoshi Furukawa	ISS	University of Tokyo, selected 1999

Recruitment groups

1	1986	Mamoru Mohri, Chiaki Mukai, Takao Doi
2	1991	Koichi Wakata
3	1997	Soichi Noguchi
4	1999	Satoshi Furukawa, Akihiko Hoshide, Naoko Sumino

Finally, in preparation for the international space station, a fourth group of astronauts was selected in 1999. Those chosen were: Atoshi Furukawa, born in Yokohama in April 1964, a surgeon in the Department of Medicine hospital attached to the University of Tokyo; Akihiko Hoshide, born in December 1968 in Setagaya-ku in Tokyo, an engineer in the Space Utilization Promotion Department at NASDA; and Naoko Sumino, the only woman of the group. She was born in December 1970 at Matsuda and was an aerospace engineer specializing in centrifuges at NASDA's Tsukuba Space Centre. They were sent for 18 months' instruction at the Tsukuba Space Centre, followed by intensified study of ISS systems in the United States. They were sent first for training in Tsukuba Space Centre in April 1999, initially in basic, lecture-based instruction, in such areas as engineering and space science. This was followed by specialized training, which covers

Soichi Noguchi

such areas as the use of space suits, spacewalking (done through underwater experience), zero gravity and survival training. Because they will be on board the station for three to six months at a time, candidates were required to pass a long-duration aptitude test.

In anticipation of an emergency return to Earth and landing a long way off course, Japanese astronauts Akihiko Hoshide and Satoshi Furukawa went to Sochi on the Black Sea for splashdown and water recovery exercises in July 1999 and were later scheduled for more rigorous winter survival training in Siberia several month later.

These preparations for the ISS represented a considerable investment by Japanese industry in general and NASDA in particular. But there was more. At the same time, Japan was experimenting with spaceplanes and important adjacent technologies for the International Space Station.

JAPANESE SPACEPLANES: THE CONCEPT

Spaceplanes date from the beginning of the space age. In 1960, the United States Air Force developed, though never flew, a manned spaceplane called the *Dyna-Soar* (short for dynamic soaring). The Soviet Union developed a plethora of spaceplane projects at the same time, like *Spiral*, but none flew into orbit with a human crew. Both countries eventually developed large space shuttles, the space transportation system or shuttle (1981) and *Buran* (1988) respectively. In the 1980s, the European Space Agency flirted with a small manned spaceplane called *Hermes* to take off on the *Ariane* rocket from French Guyana and return to a runway near Toulouse in France.

NASDA first researched the idea of spaceplanes in the early 1980s, looking at a range of different designs, such as ramjets and scramjets. In 1987, the Space Activities Commission proposed the development of a manned space shuttle, but one which should be preceded by a series of unmanned tests. Wind tunnel tests were carried out in 1988 in the National Aerospace Laboratory. Research was begun on the types of materials that could withstand reentry at temperatures of 1700°C, such as carbon-fibre-reinforced polyamide, carbon-reinforced composites, ceramic tiles and titanium.

ALFLEX flies in Australia

Following this, NASDA defined an unmanned spaceplane, HOPE, 11.5 m long to ride the forthcoming H-II rocket into orbit, in scale not unlike the manned *Hermes* project. HOPE was closely linked to the arrival of the H-II—indeed, HOPE stood for H-II Orbiting Plane. It would rendezvous with the space station, dock with the help of its manipulator system and return to a runway landing. The aspiration was that later versions might eventually be manned.

The Japanese developed their spaceplane concept in five phases:

- OREX, designed to test materials to survive through reentry, flown on the first H-II in 1994.
- HIMES (Highly Manoeuvrable Experimental Space vehicle), a 500-kg sub-orbital model designed to be launched from balloon at a height of 20 km to reach mach 4 and an altitude of 80 km, providing transonic data.

- ALFLEX (Automatic Landing Flight Experiment), a lifting body tested in the Australian desert, providing subsonic data.
- HYFLEX spaceplane, flown on the first J-1 in 1996, providing supersonic data.
- The goal was a full-scale HOPE spaceplane, unmanned at first, which would fly to the International Space Station.

Later, two intermediate programmes were added: HOPE-X, a scaled-down version of HOPE; and the HOPE Transfer Vehicle, or HTV, an unmanned can-shaped freighter shuttle.

In fact, the term 'HOPE' became a general one used to describe Japan's shuttle project and covered a range of projects associated with Japanese spaceplane development. The term has been used confusingly, covering the full range of these still evolving concepts.

JAPANESE SPACEPLANES: DEVELOPMENT

To retrace the history, precursor spaceplane development had begun in the 1980s. In 1994, a joint NASDA/Mitsubishi report proposed that Japan develop its own fleet of space shuttles. Like the NASA shuttle, the report held out the prospect of radically falling costs for access to space. Following endorsement by the Federation of Economic Organizations (Keidanren), the Science and Technology Agency and then the Space Activities Commission, HOPE first received funding in 1995. The promoters of a Japanese spaceplane argued that it would be an economical and convenient way to resupply the Japanese module at the international space station; bring back research results; supply other parts of the space station; and launch small satellites into space. In 1995, a model of HOPE was sent to the Central Aerohydrodynamic Laboratory in Russia for testing.

By this stage, progress had been made with the various precursor spaceplanes. There was an initial setback, for the first HIMES failed on 21 September 1988 when its balloon collapsed at 18 km before the mini-shuttle was even fired. The gondola separated and the spacecraft crashed into the Pacific. This was a 2-m model, weighing 185 kg, with a solid rocket motor for a mach 4 flight and gliding test. Four years later, a one-seventh model was tested in February 1992. Launched by a balloon off Kagoshima, it reached an altitude of 70 km before impacting 400 km downrange. By way of historical footnote, the rockoon concept, first advocated as far back as 1956, had now been vindicated.

ALFLEX, 9 m long, 760 kg in weight and a third the size of HOPE, was a lifting body with a squat body, dumpy nose and bent-up wings. NASDA was anxious to learn about the landing characteristics of spaceplanes, so with the cooperation of the Australian authorities, built tracking facilities in the almost deserted Woomera Missile Base in South Australia. Woomera, 500 km north of Adelaide, had been the centre for Britain's rocket efforts in the 1960s: all that was left was the airport and a minor Australian–US joint defence facility. The population had dwindled to less than a thousand people. NASDA converted a hanger to accommodate ALFLEX and its KV-107A mother helicopter, built a flight control centre and installed new navigation systems.

The first ground tests of ALFLEX were made in April 1996, suspended tests in May and the first glide test on the morning of 6th July. Towed in the air at a speed of

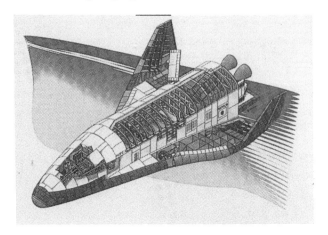

HOPE-X

130 km/h, ALFLEX was released at an altitude of 1500 m some 2.5 km from the airfield. It dived at 30°, reached a speed of over 200 km/h and then flattened out for the final stages of landing. Satellite and microwave systems guided the body in to a perfect, computerized desert runway landing.

Spaceplane precursors	
Sep 1988	HIMES (fail)
Feb 1992	HIMES
Feb 1994	OREX
Jul 1996	ALFLEX

REVIEWED AND REVISED

However, the overall development of the HOPE concept proved to be expensive. Between 1988 and 1998, ¥43bn (€383m) was spent on these precursor programmes and the subsequent designs and there was little hardware to show for it. In the late 1990s, the Japanese space programme went through a number of reviews in the wake of Japan's financial problems. At one stage, it look as if the whole spaceplane project would be abandoned and the HOPE programme would slip into the indefinite future.

Trying to save something from the reviews, HOPE-X was devised as an intermediate programme 'to establish the basic technologies for each flight phase of HOPE, using nearly the same size and configuration vehicle as HOPE ... the final stage of developing HOPE'.[9] A new, intermediate programme, HTV, was proposed for introduction in 2002.

The HOPE-X programme was for a small-scale HOPE precursor. The launch date slipped to 2000 and then to 2004, the first time a major Japanese programme had been slipped twice. HOPE-X was designed as a 12-tonne automated space shuttle able to deliver 3 tonnes to the International Space Station. It was 15 m long, with a 9.7 m wing span, a height of 5 m and had twin tails. It will be placed vertically on the H-IIA. After its mission to the station, it will head nose-first into reentry, like the American space shuttle or the Russian *Buran*, protected by its carbon and ceramic tiles, before touching down on a 1800-m runway. Its first flight will be a single orbit of 120–200 km with a touchdown on Christmas Island airfield. This will make Japan only the second country to fly a fully automated space shuttle (after Russia). Even before the maiden flight, a high-speed demonstrator will be built and tested in 2001 both by plane drop and by self-powered flight, with jets fitted at the rear.

The second intermediate programme was the HOPE Transfer Vehicle, HTV. Inspired by the unmanned *Progress* freighter vehicle which from the 1970s supplied the Soviet Union's *Salyut* and *Mir* space stations, the H-IIA will place the HTV in a 350–400-km orbit 200 000 km behind the International Space Station for a three-day chase. HTV was shaped like a cylindrical can, 10 m long and 4.4 m in diameter, weighing 15 tonnes. HTV comprised a propulsion module, an unpressurized middle section with an exposed scientific pallet and a pressurized section for astronauts to enter (in an alternative configuration, the pressurized section can incorporate the exposed section). Like *Progress*, it was strictly utilitarian and likely to prove its value quite quickly.

As it catches up with the ISS, HTV will use the global positioning system to close to 20 km from the complex. At 500 m, radar lasers will be used to bring the HTV in for the final phase, using techniques developed by *Kiku-7*. Coming up to the station from below, it will be grabbed by the manipulator arm and dock at node 2 of the ISS.

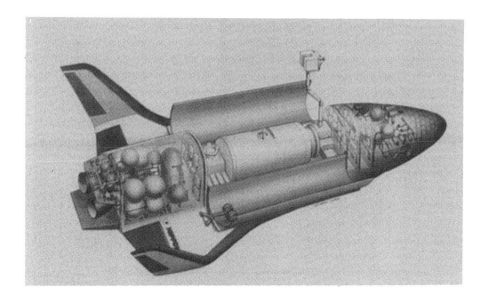

The eventual aim: HOPE

HTV will bring as much as 7 tonnes of food, clothes, water, batteries, consumables and scientific equipment up to the *Kibo* module. Two flights to the station are envisaged each year. After two weeks, it will be filled with rubbish from the station and separated for a destructive reentry. The cost of the HTV programme was estimated in 1998 at ¥27bn (€241m).

If the HOPE-X and HTV programmes prove successful, Japan will resume its onward march with the full-scale HOPE programme. In the distant future, Japanese engineers have expressed the desire to fly a fully Reusable Launch Vehicle (RLV). The idea of the RLV is that a single-stage vehicle, probably a delta-shaped shuttle, ascends to orbit, delivers payloads and returns like a plane, for a quick turnaround and relaunch. The entire vehicle is reusable and no throwaway boosters are involved. American engineers have been working on RLVs since the mid-1990s and two precursor projects were devised, the X-33 and the X-34, though both experienced problems not dissimilar to the Japanese with HOPE. A study has even been done of how the LE-5 and LE-7 engines of the H-IIA could be adapted for a Japanese RLV and aerospaceplane for the 2020 and 2030 period.

SHUTTLE LANDING FIELDS

A key problem for any shuttle project is to develop a suitable landing facility. For their space shuttle, the Americans built or modified runways at Edwards Air Force Base, California, Cape Canaveral, Florida, and White Sands, New Mexico, and made arrangements for landing fields abroad to be made available during emergencies (e.g. Rota in Spain).

Japan first considered a number of domestic candidates in the Japanese islands, such as Magejima Island, Kagoshima and Kamaishi in Iwate, but both required overflights of other countries (China and Korea respectively) and were in areas of already crowded airspace. Christmas Island, thousands of miles downrange in the Pacific, offered a good climate and already hosted a Japanese tracking station. Christmas Island had a 200 km long uninhabited peninsula, where there was an abandoned 2000 m long runway built by the British to fly in instruments for nuclear tests carried out there from 1956 to 1963. NASDA agreed to take responsibility for the development of a new runway, buildings, roads, water and electricity.

In 1999, Japan made an agreement with the Republic of Kiribati, otherwise known as Christmas Island, for a spaceport to be built on the island. The legal arrangement was for the spaceport to be leased to NASDA for 20 years. It would serve as the landing field for the HOPE-X shuttle precursor and, later, HOPE itself.

CONCLUSION: A MANNED TOEHOLD IN SPACE

By 2000, Japan had built up an experience in manned spaceflight rivalling that of Europe. There had been seven manned Japanese spaceflights—one to *Mir*, one a dedicated *Spacelab* mission, one on an international *Spacelab*, one to retrieve a free flier, a spacewalk and two shuttle missions. Four national groups of astronauts had been recruited and a manned spaceflight centre established. A spaceplane had been in design for ten years and precursor tests had been conducted. Although the HOPE spaceplane project had encountered financial difficulties and its schedule had slipped, HOPE was still in existence, unlike the Russian shuttle and the European *Hermes*. Japan therefore remained, after the United States, the only other country with a spaceplane project in the early part of the twenty-first century.

6

Japan's space infrastructure

A space programme is dependent for its success on a well-developed and well-planned network of ground, testing and communication facilities—its infrastructure. It also requires a coherent structure for planning and policy-making—what is now termed the institutional architecture. Here they are described.

HOW THE JAPANESE SPACE PROGRAMME IS ORGANIZED

The key feature of the Japanese space infrastructure from the beginning has been the division of responsibility between ISAS and NASDA, the former concentrating on small scientific missions, the latter on technology development. However, they are not the only players in the Japanese space industry, for five other government departments and agencies receive budgets for space development. About 9500 people are directly employed in space activities in Japan, tens of thousands indirectly.

The current Japanese space programme operates under the prime minister's office. Policy is determined by the Space Activities Commission (SAC), which sets the broad outline of space developments. The Space Activities Commission was set up by the Minister for Science & Technology in May 1968 and built on the previous work of the National Space Activities Council, set up in 1960. The Commission normally comprises four members or Commissioners. The Commission issued, in 1978, the first 15-year plan, *Outline of Japan's space development policy*, one which is regularly updated and is the working document for the programme. It has been renewed in 1984, 1989 and 1996. The current, 1996 plan has seven key aims: the promotion of science and technology, meeting social need, the reduction of launcher costs, the promotion of international cooperation, the development of industry, the preservation of the space environment and the balanced use of manned and unmanned systems.

Under the Space Activities Commission comes the Science and Technology Agency (STA), which is the parent body for NASDA and other institutes (e.g. National Aerospace Laboratory), but not ISAS. It has an important role in defining budgets and policies. Generally, new space programmes must have the approval of both the SAC and the STA. A third influential policy body is the Space Activities Promotion Council, set up in 1968, which is an industry body, comprising 96 companies involved in the space business.

Five key government departments and agencies take part in the Japanese space programme. These are the Ministry of International Trade and Industry (MITI), the Ministry of Posts & Telecommunications (MPT), the Ministry of Education (parent body of ISAS), the Japan Meteorological Agency and the Telecommunications Advancement Organization of Japan. The Ministry of International Trade and Industry has the largest space budget after NASDA and ISAS and has recently begun to develop its own satellite programmes (e.g. USERS). It may be expected to grow in importance.

Organization of the Japanese space programme

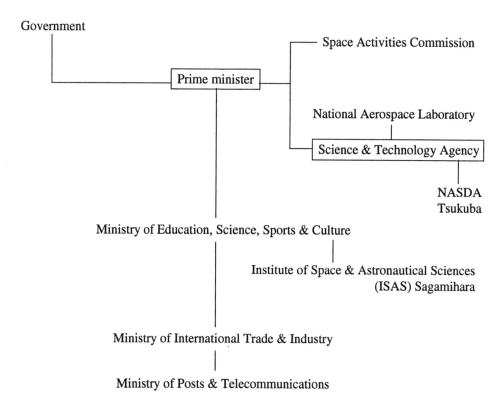

NASDA—THE NATIONAL SPACE AGENCY

NASDA is the principal Japanese space development agency, responsible for the development of the N- and H-series of rockets, engineering satellites and the manned and spaceplane programmes. NASDA has 1019 administrative personnel. Most work at its headquarters in the Tsukuba space centre and at its launch facility at Tanegashima.

> **Government ministries involved in space exploration**
> Posts & Telecommunications
> Transportation
> International Trade & Industry
> Construction
> Education, Science, Sports & Culture (ISAS)

> **NASDA divisions**
> Space Utilization
> Transportation
> Satellite Systems
> Earth Observation
> Research & Development

NASDA TSUKUBA SPACE CENTRE

Tsukuba is to the Japanese what Mission Control Houston is to the Americans and Star Town is to the Russians. Tsukuba is a science city 60 km north east of Tokyo, opened in June 1972. Seven NASDA facilities are located there. This is a tracking, control and testing centre, equipped with simulation chambers, structural test buildings and electronic test centres.

> **Main facilities in Tsukuba**
> Space Station Integration and Promotion Facility
> Space Experiment Laboratory
> Space Station Test Building
> Astronaut Training Facility
> Weightless Environment Test Building
> Space Station Operations Facility

Tsukuba hosts the main Japanese tracking facilities. Additional NASDA stations are located in Katsura (south of Tokyo), Okinawa, Masuda (near Tanegashima), Kagoshima and downrange in Chichi-Jime Island, Kwajalein Atoll, and Christmas Island. Weather data from Japanese meteorological satellites is received separately by the Japanese Meteorological Centre, set up in 1977 under the Japan Meteorological Agency, in turn responsible to the Ministry of Transport. The centre operates a ground station at Hatoyama, 35 km northwest of Tokyo to take and process signals from the GMS *Himawari* satellite series. Hatoyama has two 18-m dishes and, once signals are picked up, they are computer-processed and distributed to their wide range of users. Hatoyama is assisted by distant stations in Ishigaki Island and Crib Point in Australia.

Tsukuba has an acoustic chamber, solar simulation chamber, vibration test platforms able to shake eight-tonne payloads, a magnetic test facility and a 25 m tall thermal vacuum test chamber. There is a spacecraft clean room integration and test building for payloads on the H-II. The acoustic test facility is 17 m tall, 10.5 m in width and depth and has a volume of 1600 m³. The purpose is to anticipate the vibration of launch and the full wavebands of sound experienced in the course of a space mission, the sounds being fed in by an electro-pneumatic sound converter modulating pressurized air. The vibration test facility is able to shake satellites and modules in three dimensions (roll, pitch, yaw) to anticipate the stresses of launch.

Main NASDA space facilities

TANEGASHIMA RANGE—LAUNCH SITE BY THE OCEAN

Tanegashima, at one stage called Osaki site, is on a peninsula on the island of the same name. Around a nearby lake may be found a power plant, range control centre, telemetry station and checkout facility. The Tanegashima launch site consists of three main pad areas. These are:

- Takesaki, to the southeast, which is used for sounding rockets, H-II solid rocket booster tests and range control;
- Osaki complex, originally built for the N-I, H-I and now the J-1;

- Yoshinuba complex, now used for the H-II, with a new side pad, enabking it to handle two rockets at a time.

A launch pad called Osaki was originally built for the abandoned Q rocket, but very close to it is the pad for the N and the later H-I rocket. Considerable adaptation took place to turn the N site into the H site, involving the construction of a high mobile service tower and cold propellant storage facilities. For a launch, the rocket was prepared in a mobile service tower, before being rolled back 100 m before launch. For the final stages of the countdown, the rocket was serviced by two umbilical towers. Since then, it was adapted and simplified to take the J-1 solid rocket booster. Rockets are cradled in tall red and white gantries, their height exceeded only by a high weather tower on a hill near the Q site. In preparation for a launch, rockets and payloads are tested in the Solid Motor Test Building, the Spin Test Building, clean room and Third Stage & Spacecraft Assembly Building.

Closer to the headland is the modern Yoshinuba launch complex, set up to handle the H-II. It is one of the world's most impressive launch sites. It is less than a kilometre away from the older Osaki pad. Construction of this 150 000 m^2 site began in the mid-1980s and was completed in 1992. Yoshinuba has its own vehicle assembly building, mobile launcher, pad service tower, control centre and cryogenic storage service. Rockets are brought the short distance to the pad on the mobile launcher which rolls down an apron with parallel roads on either side and a grass centreline with the large letters 'NASDA' carved into the greenery. The vehicle assembly building is a smaller version of the famous Apollo vehicle assembly building at Cape Canaveral and is able to assemble two H rockets at the same time.

The ultramodern Yoshinuba complex

Spacecraft and rockets are put together vertically on a mobile tower in the vehicle assembly building which has a number of bays. The tower is 22 m high, weighs 800 tonnes and, once everything is ready, makes the 500-m rail journey down to the pad. Here the rocket meets a pad service tower, a fixed section with two rotating parts. This is 67 m tall, with 12 floors and uses a 20-tonne crane. Yoshinuba dwarfs the Osaki site in size. When the French remote sensing satellite SPOT flew over the Yoshinuba site during its construction, the earthworks were clearly visible, the apron resembling a short aeroplane runway. A second pad, designed to take the increased traffic likely to result from the introduction of the H-IIA, was completed in 1999. The new pad is on a slip road parallel and close to the original pad and has a new mobile service tower to bring stacked rockets to the pad.

Tanegashima also includes solid propellant storage facilities and a static firing test centre which enables the first stage to be held down on the pad for test firings of its main engines. The test stand for the LE-7 was fitted out in the course of 1989 and has a big cement deflector trench, a water coolant system and, nearby liquid hydrogen and liquid oxygen storage tanks and a high-pressure gas storage facility for helium and nitrogen. Launches at Tanegashima are controlled from blockhouses close to the pad. On the Osaki pad, there is a blockhouse 170 m from the pad, set underground. At the Yoshinuba pad, there is an underground mission control blockhouse, near the vehicle assembly building, full of consoles and computers. Not long after lift-off, control is passed to the Takesaki range control.

Yoshinuba mission control is twice the size of the Osaki centre. Shaped like a pyramidal tent, it has a control room in the centre, surrounded by telemetry and equipment rooms. Should the worst come to the worst and a rocket explode on the pad, mission controllers are protected by 1.2 m thick concrete over their heads. The air-conditioning system will provide air for a hundred controllers for four hours (there is an evacuation tunnel in any case).

Fishing restrictions historically proved to be a significant restraint on the operation of the Tanegashima launch site. However, the prospects of commercializing the H-IIA launcher depended on a more flexible and longer launching season. Accordingly, new agreements were reached between the government and the fishermen's union to extend the launch window from 90 days to 180 days, while still preserving the prime fishing period of March to mid-June.

Launching and fishing season in Japan (Tanegashima)

Old system		New system	
February–September	90 days	Jan–Feb Mid-June to mid-Sep Nov–Dec	180 days

The main centre for the testing of rockets in Japan is the Kakuda Propulsion Centre, established July 1980, where all Japan's high-performance liquid-fuelled rockets were produced. Here, the LE-5 and LE-7 series of engines were perfected. The centre has a high-altitude test firing stand to simulate firings in the space environment and a tank to simulate vacuum, heating and solar radiation.

ISAS—HOME OF SCIENTIFIC SPACE PROGRAMMES

ISAS was the organization originally involved in space development in Japan and has since built up an extensive range of facilities. Its headquarters are located in Sagamihara, 40 km west of Tokyo and facing mountains in Tanzawa. The centre comprises a mixture of high- and medium-rise complexes, not unlike a modern university campus. ISAS has nine research divisions and seven centres. The research divisions are astrophysics, plasma, planetary science, basic space science, systems engineering, transportation, propulsion,

---

ISAS main facilities
Launch site: Kagoshima, 1962
Noshiro Test Centre, 1961 (solid and liquid-fuelled rockets)
Usuda Deep Space Centre
Sanriku Balloon Centre
Space Utilization Research Centre, Sagamihara, 1988
Centre for Planning & Information Systems, Sagamihara, 1993
Centre for Advanced Spacecraft Technology, Sagamihara, 1995

ISAS space facilities

spacecraft engineering and applications. Its centres are the Kagoshima launch centre, Nishiro testing centre, Sanriku balloon centre, Usuda deep space centre, the space utilization research centre, the centre for planning and information systems and the centre for advanced spacecraft technology. ISAS has 324 academic staff and technicians with an additional 174 students on its books.

Noshiro is located at Asani beach in the north-west of the main island of Japan, Honshu, away from population centres. It is used for testing solid and liquid fuel rocket motors including cryogenic engines and has vertical and horizontal test stands and fuel tanks. Usuda Deep Space Centre is right in the middle of the island of Honshu. It is high up in the mountains, 1450 m above sea level. Its original 64-m dish has now been joined by a smaller 10-m antenna. The main missions for Usuda have been *Sakigake*, *Suisei*, *Voyager 2* and *Haruka*.

The Space Utilization Research Centre is the main location for ground tests of possible future space experiments. The role of the Centre for Advanced Spacecraft Technology (CAST) is to pioneer the next generation of spacecraft technology to be used in lunar, planetary and astronomical missions. The purpose of the Centre for Planning and Information is to process, store analyse and distribute space science data, which is done through a range of databases, mainframe and supercomputers. Sanriku Balloon Centre, completed 1971 near Mount Ohkubo on the Pacific coast, launches balloons into the atmosphere where they can measure the weather and carry out astronomical observations. 230 balloons have been launched since it opened, averaging about 15 take-offs a year. Some balloons have flown as high as 46 km and others have reached China.

ISAS KAGOSHIMA LAUNCH CENTRE

Kagoshima is Japan's first launch facility, used to fire the solid rocket boosters of the ISAS scientific programme. It is built on plateaus and in valleys at a hilly mountain site in southern Japan. Inland lie the control, tracking, and payload integration centres. Nearer to the coast is the area used for the launch of the first Japanese satellite, called the *Lambda* Centre. Closer to the coast is the *Mu* centre, comprising an assembly building, control tower and launch stand. Satellite preparation buildings and administrative complexes complete the launch site.

The solid rockets fired from this range are erected by their tower and then tilted toward the sea at the angle at which they should be fired. Normally rockets streak at an angle of 80° out over the sea as they speed toward their orbital and planetary destinations. Optical trackers are located 2 km away on the Miyabaru plateau.

KEY COMPANIES

As is the case in Russia, Europe and the United States, the production of spacecraft, rockets and equipment is contracted out to large industrial, defence and technology contractors. Many of the Japanese companies are household names in the west because of the wide range of their products, such as cars and electronics.

Noshiro testing centre

The principal industrial companies involved in the Japanese space programme are Ishikawajima Heavy Industries (IHI) (upper stages and small engines), Mitsubishi (liquid-fuelled rockets), Nissan (solid-fuelled rockets) and Rocket Systems Corporation (marketing of launch services, e.g. the H-II).

Mitsubishi had a long involvement with the N series and the H-I and was prime contractor for the H-II, building both the rocket bodies and the engines. Mitsubishi has its own test stands and facilities (e.g. Tashiro Test Field, Nagoya). Nissan makes the *Mu-5* satellite launcher and many of Japan's other solid-fuel rockets and strap-on boosters. Nissan has its own range of manufacturing and test facilities—a research and development centre at Kawagoe and a test facility at Taketoyo where solid rocket engines may be static tested. For satellites, the main Japanese industrial companies are Mitsubishi Electric (Melco), Toshiba, Kawasaki and Nippon Electric Company (NEC). Many of these companies are large global trading companies with substantial in-house research and development experience. For example, Nissan has a large 480 000 m^2 plant in Tomioka with 800 staff devoted to the design, development, production and testing of space equipment.

Currently, the main companies involved in the provision of satellite-based television services are Space Communications Corporation of Japan, Japan Satellite Systems, the Broadcasting Satellite System Corporation and NHK (Nippon Koso Kyokai) and these have in turn a long history of inter-relationship, merger and rivalry.

JAPANESE SPACE BUDGET

Space spending by Japan got off to a slow start, only about ¥2.940bn (€26.2m) being committed a year in the mid-1960s (1966 figure). Spending began to rise in the mid-1970s, associated with the costs of developing the N rocket, but was still well behind the space budgets of France or West Germany.

Japan's space budget in 1999 was ¥251bn (€2.241bn), up 1.46% on the previous year. This was subdivided into ¥182bn (€1.25bn) for NASDA (72.8%) and ¥19bn (€132m) for ISAS (7.73%), the balance going to the other government departments and agencies involved in space exploration. The current Japanese space budget makes it the fourth spacefaring nation, after US, Russia and Europe (bearing in mind that realistic comparative figures are not available in the case of Russia).

The figure shows the relatively small proportion of spending devoted to ISAS, the steady growth of the NASDA budget and the development of a more substantial 'others' budget since the late 1980s. The dip in spending in the late 1980s, and the plateauing of spending in the late 1990s, are also evident. The diversification of the budget is evident in the second figure. The Ministry of Transport (MOT) now receives a larger budget than ISAS and a significant budget now also goes to the Ministry of International Trade and Industry (MITI).

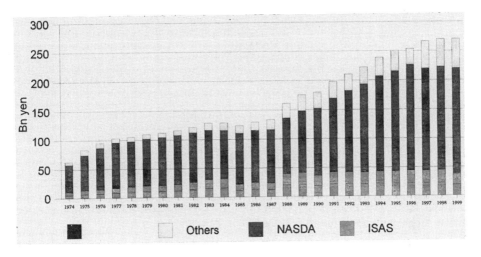

Japanese space budget, 1974–99, bn yen

Within the NASDA budget, the largest amounts are for the space station and, within that, for the *Kibo* module. Examining it by sector, 27% goes on launchers, principally the updating of the H-II and the J-1. 21% goes on the international space station, 17% on observation of the Earth and 15% on the construction of satellites. The balance, 20%, goes on research and development, information, and administration.

CONCLUSIONS: THE FOURTH SPACEFARING NATION

Japan has a solid spaceflight infrastructure and is the fourth spacefaring nation in the world, after the United States, Russia and Europe. The country has the full range of launch sites, test facilities, tracking centres and space development institutes necessary to sustain a long-term programme well into the twenty-first century.

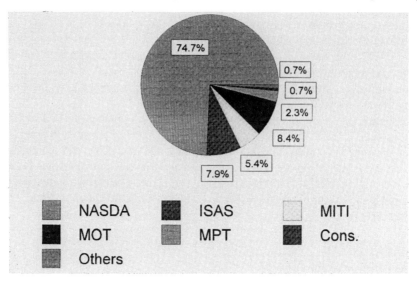

Division of Japanese space budget, 1999, bn yen in percentages

7

Future prospects

By 2000, the Japanese space programme had achieved notable successes. A strong space industrial infrastructure had been built and the country had thirty years' experience in building rocket launchers and sending satellites into Earth orbit. Probes had been sent to the Moon, comets and Mars. An advanced indigenous rocket had been built and new prospects for manned flight were coming to fruition with the construction of the International Space Station from 1998. What prospects are in store for the Japanese space programme in the twenty-first century?

Three deep space missions are in prospect and these are reviewed here. They are to the Moon and the asteroids.

LUNAR A: TO THE CORE OF THE MOON'S SECRETS

Plans for a wide-ranging programme for lunar, planetary and cometary exploration were first articulated by ISAS in the late 1980s. Following the success of *Hiten*, ISAS proposed a second lunar probe called Lunar A, but this would be a scientific rather than an engineering test mission. ISAS's move was timely, for it coincided with a revived scientific interest in the Moon and the discovery of water there in the late 1990s by the American probe *Lunar Prospector*.

The aim of ¥10.5bn (€93.75m) Lunar A mission was to drop two 13-kg penetrator probes into the lunar surface. Shaped like torpedoes, 15 cm in diameter and 83 cm long, each would carry a heat flow probe and a seismometer to measure moonquakes, thereby solving the mystery of the Moon's core. Heat flow experiments had been carried by Apollo, but the Japanese ones were five times more sensitive. The heat flow experiment was designed to detect whether any heat was still emanating from the Moon's core, how much, and what the size of the core was. The Apollo seismometers detected three types of moonquakes—shallow ones from the impact of meteorites, those sparked off by Earth's tidal forces and others coming from deep in the Moon' core. Lunar A would improve on this information.

Although the idea of a penetrator sounded simple enough, designing equipment to survive a 300 m/sec impact that caused up to 10 000g was no mean task (g = the force of Earth's normal gravity). By the early 1990s, Japanese scientists had tested several

Lunar A: the penetrators are visible at both sides

penetrators, firing models from a nitrogen gun into sand boxes designed to resemble the lunar surface. On 17th September 1995, a S-520 sounding rocket was launched to test the separation method for the orbiter and the penetrators, but the mission went awry and telemetry was lost after 25 sec.

The main spacecraft was a 550-kg orbiter, 2.2 m in diameter, 2 m high with a single instrument, a 20-m resolution camera. Following launch, Lunar A will follow a gentle curving trajectory to 340 000 km near to the Moon taking two months. It will pass the Moon and, using four solid rocket motors to brake speed by 550 m/sec, nudge itself into a 200 km Moon orbit on the second near-lunar pass four months after launch.

The penetrators will be released a month after insertion. The release of the penetrators in lunar orbit will be a critical moment. The orbiter will descend to 40 km one month after lunar orbit insertion for this delicate manoeuvre. Once the penetrators are released, a motor fires to kill forward motion. Jet thrusters then turn the penetrators to the vertical, so that they free fall at as slow a speed and as vertically as possible to the lunar surface (they must come in at not more than 8° from vertical). The aim of Lunar A was to deploy one penetrator 45 km north of the Apollo 14 landing site of Fra Mauro and the second on the

Japan–future deep space plans		
Lunar A	2002	Two 13 kg lunar penetrators/seismic stations on far side, Apollo 14 site. Spacecraft: 550 kg. Aim: find out if Moon has a molten iron core, measure heat flows.
Muses C	2002	Asteroid 1989 ML. Arrive 2003; return sample in 2006.
Selene	2004	Lunar orbiter, 100 km, & lander. Biggest lunar orbiting package since 1970s. *Selene II* in preparation.

lunar far side in the middle of the relatively flat crater. It is expected that the penetrators will bury themselves up to 3 m into the Moon at a speed of 285 m/sec at a rate of 10 000g. Half of the penetrator is motor and half is instrumented with seismometer, battery and transmitter.

After deploying the penetrators, the mother craft will at as a data relay and provide television on its lunar imaging camera. Every 15 days it will pass over the penetrators and unload the data from the penetrators to Usuda Deep Space Centre.

Designing the separation process has proved to be extremely problematical. Serious question marks were raised during design reviews in 1997 and the project put on hold for two years while these were sorted out. Then the recalculation of orbital mechanics found that a 1999 launch would leave Lunar A in an unlit, power-starved orbital path with unusually long periods of darkness. The outcome of a stormy design review was to cut the number of penetrators from three to two, substituting the third with extra battery power.

More trouble was to follow. Further penetrator tests suggested the danger of a wire breaking between the seismometer and the data relay system and that the design had insufficient protection against battery leaks after the violent impact. These threatened yet more delays. The launch was delayed again for another two years to 2002.

SELENE'S SOFT-LANDING

Following Lunar A is *Selene*, which stands for Selenological and Engineering Explorer, a ¥21bn (€188m) project for 2004 led by Nippon Electric to send a large, 2.8-tonne payload into lunar orbit with 300 kg of scientific instruments and a 350-kg lander. This is a joint ISAS and NASDA project and will use NASDA's H-IIA launcher. This required the formation of a 300-strong team in 1996, causing much agonizing about how to reconcile the different approaches of ISAS and NASDA. The solution agreed was that ISAS would handle the scientific end of the project and NASDA the engineering aspects. The project went through a number of design evolutions, including cost-cutting measures in the late 1990s.

***Selene's* instruments**
X-ray spectrometer
Gamma-ray spectrometer
Multi-band imager
Spectral profiler
Terrain camera
Radar sounder
Laser altimeter
Radio source for very long baseline inferometry
Magnetometer
Plasma imager
Charged particle spectrometer
Plasma analyser

After a 127-hr lunar coast, the 3-tonne *Selene* mother spacecraft will be put into a 100 by 13 000 km elliptical polar orbit, one which will be adjusted to 100 by 2400 km. On arrival, the box-shaped satellite will extend 15 m long sounding antennae and 12 m long magnetic booms. A 40-kg drum-shaped relay and gravity field subsatellite will then be released for a 14-month mission before the orbiter is lowered to a 100-km, 2-hr circular orbit. The relay satellite will transmit signals from the orbiter when it is around the far side and help to measure the Moon's gravity field.

Finally, in a 100-km, 2-hr, 95° mapping orbit, a year's work of lunar orbit studies will begin using 13 instruments including a radar sounder to map lunar geology to a depth of

Selene Moon probe

5 km, X-ray and gamma-ray spectrometers to identify the composition of the lunar surface, a multispectral imaging system to assess minerals, laser altimeter to measure height, and a high-resolution stereo mapping camera. It will be the biggest payload in Moon orbit since the American Apollo and Russian Luna series in the 1970s. After a year, a 420-kg lander will separate. Shaped like a multisided box with four landing legs, 2.2 m in diameter and 1.1 m high, it will come down from orbit in a 28-min powered descent like the American lunar lander at a site yet to be decided. It will carry a camera to survey the lunar terrain for two hours after landing and then transmit further information for two months.

Selene should be the biggest boost to our detailed knowledge of the Moon since the 1970s. It carries a formidable pack of instruments, much more sophisticated than the small American *Clementine* and *Lunar Prospector* mission of the 1990s. *Selene* will make possible detailed geological maps down to 5 km below the surface.

In 1999, NASDA and ISAS began to consider a *Selene II* mission and held a planning symposium. An even more ambitious mission is proposed, using a 3.7-tonne spacecraft with seven 50-kg penetrators (four nearside, three farside) with seismometers and heat flow meters, a 100-kg astronomical telescope to be landed near the lunar pole and a 60-kg lunar rover able to travel 25 km carrying a mass spectrometer and a gamma-ray spectrometer. If successful, the mission would add a further quantum amount to the scientific knowledge of the Moon, especially its material elements, mineral composition, topography, geological structure and gravity field.

RENDEZVOUS WITH AN ASTEROID: MUSES C

Muses C is a 400 kg interplanetary probe due to launch in July 2002 for sample return mission to an asteroid using electric xenon propulsion. This is a complex mission, involving closing in on near-Earth asteroid 1989ML, landing on the surface (October 2003), sampling rocks, enclosing them in a container and sending them safely back to Earth for recovery in a tiny 18-kg cabin (June 2006). 1989ML is a typical black chondrite asteroid, so it is hoped that the sample will tell us much about the class as a whole. An engineering model is set to test out aspects of Muses C on the 2001 first flight of the H-IIA. ISAS emphasizes that the mission is more about testing new engineering technologies than about science. The mission follows in the footsteps of America's Deep Space 1, which in the later 1990s tested ion engines, autonomous navigation systems (or 'autonav' for short) and remote rendezvous.

Muses series		
A	24 Jan 1990	*Hiten/Hagoroma* lunar probe
B	12 Feb 1997	*Haruka* Very Long Baseline Interferometry
C	July 2002	Asteroid 1989ML (scheduled)

After leaving Earth on a *Mu-5* rocket, Muses C will swing outward across the solar system to intercept the asteroid for a 70-day-long interception guided by an autonomous navigation system. The trajectory will be adjusted by four ion engines which will fire for 18 000 hr altogether. The interception of the asteroid will be the most difficult phase of the mission. The space probe will be over 250 million kilometres away from Earth at the time, so it is at this point that the autonav system will come into its own. Above all, autonav must ensure that there is no collision. The autonav will steer Muses to a hovering point 20 km away while it tries to calibrate what it sees with its built-in terrain recognition system. Muses will then close in, ever so slowly at 10 cm/sec, before the close encounter beings.

Asteroids are so small and have such a weak gravity that it is virtually impossible to land on them in the conventional sense, so Muses will attach itself to the asteroid like a spider. A projector will bombard the surface, causing fragments of the asteroid to bounce back into a fragment catcher, from which samples will be assembled and transferred into a return capsule. A tiny 1-kg nanorover, the size of a child's toy, will be put down on the surface. After sampling is done, the return cabin will be detached and sent on a rapid return trajectory to the Earth. The cabin will slam into Earth's atmosphere at 8 km/sec. NASA is providing technical support for two key aspects of the mission—the landing on the asteroid and the heat shield for the return to Earth.

Luna A, *Selene* and Muses C represent the current Japanese plans for deep space exploration. However, even as they were preparing them, the Japanese faced a threat closer to home.

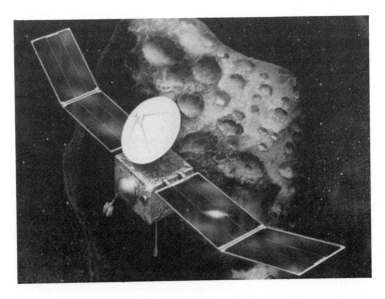

Muses closes in on asteroid

SPY SATELLITES: THREAT ACROSS THE SEA OF JAPAN

The early space programmes of the big powers had a strong military orientation—such as the *Corona* programme in the United States and *Zenith* in the Soviet Union. By contrast, the Japanese programme was entirely civilian, reflecting the postwar and constitutional settlement which restricted Japanese defence spending and prevented the country from building up armed services (though a limited 'self-defence force' was permitted).

This changed, following the strange events of 31st August 1998. Across the Sea of Japan, North Korea converted a military rocket, set it up, attached a small solid rocket motor and launched it from Musadan-ri in north Hamgyong on its east coast. North Korea hailed what it claimed was its first satellite to be put in orbit. The small sputnik, with a weight estimated at 50 kg, was called *Kwangmyongsong* or 'bright star', carrying a transmitter broadcasting 'the immortal revolutionary hymn of General Kim Il Song' and the 'Song of General Kim Jonge'. North Korea called the long, thin launcher the *Taepo Dong 1* and announced the orbit to be 219 by 6978 km, 41°, 165 mins. However, the launch was not tracked, neither at the time nor subsequently by United States Space Command, nor by the Russian space tracking system. No one outside Korea ever saw the satellite or heard the immortal hymns being relayed to Earth for their benefit. A year later, the North Koreans still adamantly insisted that it had broadcast for nine days and would orbit the Earth for another year. The American intelligence community appears to have been caught napping and originally ridiculed the North Korean claims to have launched a satellite. More mature consideration led to the conclusion that a serious satellite attempt had been made, but that it had probably not succeeded. Later, it appeared that the flames of the ascending launcher had been spotted by an American reconnaissance satellite, but the information had not been passed on.

For the Japanese, the launch of *Kwangmyongsong* had sinister implications. The Japanese noted that the ascent path of the *Taepo Dong* curved over the northern Japanese islands, bringing with it the danger of rocket stages or débris falling to Earth (parts did indeed fall near Sanriku in northern Japan). More to the point, it meant that Japan was now unmistakeably within missile range of North Korea. In its report on the incident at the end of October 1998, the Japanese Defence Agency argued that North Korea's intentions were to test military missiles rather than put a satellite into orbit. For the Japanese, the *Kwangmyongsong 1* emphasized their dependence on American intelligence. The United States had a huge electronic signal intelligence ('sigint' in the trade) eavesdropping facility at Misawa on the northeast tip of Honshu and up to twenty advanced spy satellites in orbit at any time. Whilst this system could undoubtedly keep a close watch on the North Koreans, the real question was the preparedness of the Americans to pass on information and warnings in a timely manner, something they clearly had not done.

Japan responded initially by vocal diplomatic protests. The cabinet discussed the launch the following day. The Japanese parliament, the Diet, was convened. Mitsubishi Electric proposed a reconnaissance satellite programme. Several plans were considered: a multi-purpose Earth observation programme with military capabilities; the conversion of the proposed Advanced Land Observation Satellite (ALOS) to part-military use; and, the option chosen, a dedicated military reconnaissance programme. The political outcome was approval for a programme of spy satellites to effectively monitor North Korean rocket developments that might in the future threaten Japan. Within months, the government of prime minister Keizo Obuchi had drafted plans for a ¥200bn (€1.7bn) programme for up to four 85° polar-orbiting spacecraft, 850 kg each, to be launched by

USERS

2002, two to be radar-based and able to see through cloud and darkness; and two to be optically based (resolution 1 m), circling the Earth at 500 km. The spacecraft were expected to carry heat-sensing detectors to spot the flame of a rising rocket. There was pressure on the government to buy in the spy satellites directly from the United States to save time, but the administration's view was that the technology should be developed domestically: it was expected that their data would be supplemented by that of the ALOS spacecraft. In 1999 Mitsubishi Electric (Melco) was selected to carry out preliminary work on the spy satellite' optical sensor, synthetic aperture radar system and manoeuvring systems. The Ministry of Transport was put in charge of the programme, rather than NASDA. The launcher will be the H-IIA. Japan also put money into anti-missile defence research.

Eventually, in late 1999, it was agreed that Japan would buy in a number of key parts of the reconnaissance satellite project from the United States—principally data recording and interpretation systems and the mechanism for controlling the optical sensors. The rest of the system would be built at home. A cooperation agreement between Japan and the United States was duly signed, Japan taking the view that without American cooperation the 2003 deadline would not be met.

ENGINEERING: A NEW SET OF NAMES—USERS, SERVIS, MDS

Engineering and technological development has been a theme of the Japanese space programme ever since the introduction of the *Kiku* series. Several new missions are now in prospect: USERS, SERVIS and MDS.

In the late 1990s NASDA began to prepare for launch in 2000 the first of two small, 450-kg satellites called MDS. MDS stand for Mission Demonstration Satellite. The first will be placed in a geostationary transfer orbit to study the influence of the space environment on the Earth, especially radiation; the second in a circular orbit at 550 km to study particles in the atmosphere with a 250-kg lidar.

At the same time, the Ministry for International Trade and Industry, MITI, provided additional resources for engineering tests. Two main spacecraft are involved: USERS and SERVIS. USERS, or the Unmanned Space Experiment Recovery System is a large, ¥31.698bn (€283m) project for a 800-kg service module carrying technological experiments and a 900-kg reentry module which will bring the results of the first round of experiments back to Earth. Launch is set for H-IIA in 2002 for a 500-km, 30.5° orbit. The spacecraft has 15.5-m solar arrays. USERS is intended to grow crystals for six months before they are returned to Earth (the Superconductor Gradient Heating Furnace). There is a reentry module in the form of a 1.6 m diameter cone-shaped cabin which will propel itself away from the box-and-wings service module, fire a small solid-fuel retrorocket and reenter. The experimental results will be well protected in the cabin, which will come down in the ocean near the Ogaswara Islands, while the mother spacecraft goes on to complete a three-year mission.

SERVIS stands for Space Environment Reliability Verification of Integrated Systems. Two SERVIS spacecraft will succeed USERS in 2003 and 2007. They have been funded but the design is still in definition. They will be used to test technology as diverse as car design, components for future space systems and new global positioning systems.

NASDA put forward its first proposals in 1998 for small satellites, following a global trend toward 'smaller, faster, cheaper' missions first sloganized by 1990s NASA chief Daniel Goldin. Japan was in a good position to develop such missions, having used miniaturization in its satellite designs more than most other nations. Small satellites, suggested NASDA, would be prototypically around 50 kg in weight, cuboid in shape and used in Earth orbit or deep space missions. Their development would place a premium on miniaturization, automation, autonomous navigation, attitude control systems and battery power. The first proposal for a specific spacecraft emerged later that year with a NASDA plan for a 50-cm, 50-kg piggyback hexagonal satellite designed to test out space computers and tracking systems.

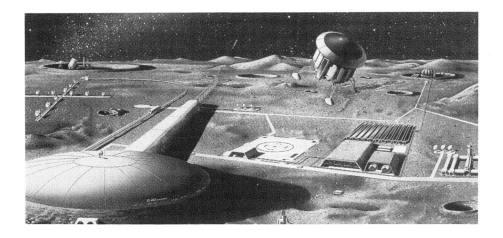

Future Japanese moonbase design

Manifest of Japanese launches
2001
ADEOS 2/WEOS   H-II

2002
Lunar A   *Mu-5*
OICETS   J-1
DRTS-W/MDS   H-IIA
Artemis/DASH   H-IIA
Astro F   *Mu-5*
Muses C   *Mu-5*
USERS   H-IIA
Test HTV   H-IIA
DRTS-E   H-IIA
ETS-8   H-IIA
ALOS   H-IIA

2003
HTV-1   H-IIA
SERVIS   H-IIA
HTV-2   H-IIA
Astro F   *Mu-5*

2004
Solar B   *Mu-5*
Selene   H-IIA
HOPE-X   H-IIA

2007
SERVIS II   H-IIA

Launches by other countries
Sep 2002   JEM-1/*Kibo*   Shuttle
2003   JEM-2   Shuttle
2003   JEM-3   Shuttle

WHALES

One of the more unusual space projects approved in the Japanese space budget for 1999 was one to track whale. ¥42m (€377,000) was allocated. The idea was originally presented by Dr Takeshi Ori of the NEC Corporation and Dr Tomonao Hayashi of the Chiba Institute in 1995. The proposal is to use satellites to monitor the migration of the blue whale, the largest and most endangered species of whale. The whale would be harpooned and, attached to the harpoon would be a 10-kg float which would have a satellite transmitter sending reports beamed up to a 50-kg Whale Ecology Observation Satellite (WEOS) which would fly piggyback on a larger payload, ADEOS 2, on the H-IIA. Electric power would come from a system developed by Seiko watches which would recharge the battery in the float from the whale's swimming motion. The satellite will pick up data from a number of blue whales, telling of their movement, the water temperature, the periods of time spent by the whale on and under water, sending the data down in a daily communication pass.

LONG-TERM AMBITIONS

This is for the short term. Japan has not neglected to think ahead and contemplate its future in space exploration. It has always been conscious of the need for long-term planning, hence the early start on space shuttle designs in the 1980s and work on a national aerospaceplane by the National Aerospace Laboratory in the 1990s (330 staff were assigned to the project at one stage). Mitsubishi Heavy Industries was involved in a series of studies of a Concorde-shaped aerospaceplane 94 m long weighing 350 tonnes which would take off and land like a conventional jet but once in space fly a mission similar to the space shuttle. The body of the aerospaceplane would contain a giant fuel tank of slush hydrogen and it would fly on a mixture of rocket and scramjets.

The Space Activities Commission Task Force for Long-term Space Policies made a far-reaching report in July 1994, in effect a revision of the original 15-year space plan of 1978. It took an optimistic view of Japanese space progress, clearly encouraged by the recent success of launching the H-II. Its core proposal was that Japan should double its space budget over the next 15 years, increase its commitment to interplanetary missions and become a leader in Earth observation. The commission proposed that in addition to its participation in the American space station, Japan would send rovers to Mars and bring back rocks from Mercury and the Moon. Over the longer period of 30 years, the commission proposed the development by Japan of unmanned and then manned space shuttles, a Japanese-led space station and ultimately a manned astronomical outpost on the Moon.

NASDA published a number of paper studies for lunar missions in the early twenty-first century. These included orbiters, landers, rovers and sample return probes to lay the groundwork for what would ultimately be a lunar base. Paper studies have been done of a lunar rover project, a sample return mission and a Venus–Mercury flyby. Model lunar rovers have even been tested. Workshops on lunar exploration, manned and unmanned, have been held at Tsukuba space centre since the 1980s, leading to sketches of Moon bases.

Private industry has also circulated proposals. In 1986, the Shimazu corporation formed a space project office and announced plans for *Lunar city 2050*—a south lunar pole city of 10 000 inhabitants and for Earth orbiting hotels. Kawasaki has published designs of a fully reusable single-stage-to-orbit rocket called the *Kankoh-maru*. This is a stubby, headlight-shaped 550-tonne rocket propelled by liquid hydrogen and liq d oxygen built to bring 50 tourists into Earth orbit for flights of between one orbit and a day in space. A similar concept was first proposed by the Japanese Rocket Society, established in 1956.

CONCLUSIONS

Japan has a series of challenging space missions planned for early in the twenty-first century. The two planned Moon probes, Lunar A and *Selene*, along with the Muses C mission to the asteroid belt, are likely to attract the most public attention. However, missions closer to home such as *Artemis* and USERS will be important for the continued development of cutting-edge space-based industrial technology. Much further into the future, plans for Earth-orbiting space hotels and lunar polar bases appear to be fantasy now. However, most space projects can be traced far back to antecedents that seemed improbable at best when they were first put forward. Previous example suggests that countries which lay a firm groundwork have, ultimately, a good chance of realizing their ambitions.

Part II
India

8

Origins

The ancient texts of Sanskrit—the *Vaimanika Shastra, Rmanyane, Rig Veda* and *Mahbharatha*—tell of how, many years ago, the whole sky blazed when a rocket ship called the vimana ascended into the heavens. The best translation of the Sanskrit tells us that the vimana 'gave forth a fierce glow, the whole sky was ablaze, it made a roaring like thunderclouds' and took off.

Interpreting the fragments of ancient Sanskrit is a dangerous exercise. These texts have received as many interpretations as there are interpreters. They range from those who see them alternatively as a rocket user manual, to an Indian equivalent of the Bible's Book of Revelations, to an early New Age fantasy, to clear proof of the landing (and subsequent return) of an alien civilization from another galaxy. Either way, they are the earliest written account of anything approximating to a rocket [10].

India's history of astronomy dates to ancient times, reaching its peak in the fifth century under the great astronomer and mathematician Aryabhata. Observatories were established. Bamboo-and-iron rockets were used against British colonists in Srirangapatna in 1792. Waves of invasion and colonization led to the decline of science in India and it was not until independence in 1947 that India was able to make a fresh start. The concept of harnessing science to build the modern India was in many ways the achievement of modern India's first prime minister, Jawaharlal Nehru, who declared:

> Science alone can solve the problems of hunger and poverty, insanitation and illiteracy, of superstition and deadening custom and traditions, of vast resources running to waste, of a rich country inhabited by starving people.

FATHER OF INDIAN ASTRONAUTICS, DR VIKRAM SARABHAI

The Indian space programme was founded by Dr Vikram Sarabhai under the aegis of the Department of Atomic Energy. He was appointed chairman of the Indian National Committee for Space Research which was charged by the government with responsibility for organizing space research in the country. In 1962, he set up India's first space base in the small fishing village of Thumba near Trivandrum in Kerala. He initiated a programme for the Indian manufacturing of French-built *Centaure* sounding rockets to be launched from there. 21st November 1963, a date which now marks the opening of the Indian space programme, saw the launch of a small, 23.3-kg American *Nike-Apache* sounding rocket from Thumba to a height of 200 km. The launch centre later became known as the Thumba Equatorial Rocket Launching Stations (TERLS). In due course the *Centaure* rockets were regularly fired from there and so, later, were Indian rockets.

Vikram Sarabhai (1919–71) was born on 12th August 1919 in Ahmedabad into a well-off family and educated at the family school directed by his mother, Saraladevi Sarabhai, a progressive and child-centred school. For higher studies he went to Gujarat College in Ahmedabad. He studied in Britain, taking a degree in natural science in St John's College, Cambridge, in 1939. Returning to India, he worked on cosmic ray physics at the Indian Institute of Science, Bangalore, under Sir C.V. Raman. Once the war ended, he returned to Cambridge and received his doctorate in photofission from the Cavendish Laboratory in 1947 (it was called 'Cosmic ray investigation in tropical latitudes'). Back again in India, he set up the Physical Research Laboratory to study cosmic rays and atmospheric physics, working there till 1965.

He also became involved in a range of industrial activities. He managed the Swastic Oil Mills in Bombay, which made cosmetics and detergents, and Standard Pharmaceuticals in Calcutta, where he introduced the industrial manufacture of penicillin. He set up many new enterprises, such as Sarabhai Chemicals, Sarabhai Glass, Suhrid Geigy Ltd, Sarabhai Engineering and the medical research group, the Sarabhai Research Centre, Baroda. To develop the textile industry, he established the Ahmedabad Textile Industry Research Association (1956). To improve management standards in India, he set up the Indian Institute of Management (1962). He founded the Community Science Centre to promote science education for children and assisted in the establishment of the Nehru Foundation to promote ideas on development. He co-founded the Darpana Academy of the Performing Arts. Well might one of his successors in the Indian space programme, Professor U.R. Rao, later describe him as 'a great institution-builder'.

FOUNDATION

Under Vikram Sarabhai's direction, India launched its first home-build sounding rocket from Thumba in 1967. It was just 75 mm in diameter, the *Rohini 75*. This was the first in a family of Indian sounding rockets, from the 2.3 m tall *Rohini 125*, to the 4 m *Menaka* to the 10 m tall *Rohini 560B*. From 1969, Indian-made propellants were used.

In 1966, Sarabhai was promoted to chairman of the Atomic Energy Commission and Secretary to the Department of Atomic Energy. The department had a broad brief which included space technology and it was there that Vikram Sarabhai devised a first development plan for the Indian space programme, one which emphasized the use of innovative technologies from around the world designed for India's benefit, the use of low Earth orbit satellite monitoring for agriculture and forestry, and the use of satellite communications to educate children in remote villages.

It was at this stage that the path of Vikram Sarabhai crossed with that of the father of the Japanese space programme, Hideo Itokawa. Both met at an international astronautical conference and became personal friends. Sarabhai persuaded Indira Gandhi to invite Itokawa, then in the process of leaving the Japanese space programme, to be adviser to the Indian space programme for three years (1965–67). Sarabhai and Itokawa worked

Vikram Sarabhai

together under the guidance of the great Indian physicist Dr Homi Bhabha, who until his sudden death in a plane crash was chairman of the Atomic Energy Commission which then guided all scientific and applied research in India. Itokawa regularly travelled with Sarabhai between Delhi, Bombay, Ahmedabad, Madras and Trivandrum, supervising the setting up of the institutes of the Indian space programme. He found Sarabhai a hard man to keep up with, for the Indian slept only three hours a day, so relentless was his pace of organization. It was Hideo Itokawa who proposed to Sarabhai the name *Rohini*, one of the Hindu gods, for India's rockets. Indian scientists visited Japan in 1966 to study the progress of the University of Tokyo. At the end of his life, Hideo Itokawa wrote of this period in his 1994 book, *Third road—India, Japan and entropy.*

Vikram Sarabhai set himself resolutely against the use of space programmes for prestige purposes, as some form of national virility symbol. Instead, they had to be harnessed for the immediate, practical needs of the people, an ever-greater imperative in the case of a developing country:

There are some who question the relevance of space activities in a developing nation. To us, there is no ambiguity of purpose. We do not have the fantasy of competing with the economically advanced nations in the exploration of the Moon or the planets or manned space flight. But we are convinced that if we are to play a meaningful role nationally and in the community of nations, we must be second to none in the application of advanced technologies to the real problems of man and society which we find in our country. The application of sophisticated technologies and methods of analysis of our problems is not to be confused with embarking on grandiose schemes whose primary impact is for show rather than for progress measured in hard economic and social terms.

Vikram Sarabhai was obsessed with the provety of the Indian rural areas and the widening gap between the developed nations and the emerging developing countries. He argued that poor nations did not have the luxury of building up their industry and technology step by step. Rather, they must leapfrog to advanced technology and harness it for their specific needs, not for prestige but for sound economic reasons.

Under the guidance of Vikram Sarabhai, the programme adopted two formal aims:

- The rapid development of mass communications and education, especially for the remote rural areas
- Surveying and management of the country's natural resources

In 1964, American television satellites had transmitted the Olympic Games live from Tokyo. This impressed Indians greatly and spurred the setting up of satellite receiving stations in the country. The initiative for Earth stations was taken by Vikram Sarabhai in the Department of Atomic Power in 1964. He brought together a team of engineers at Ahmedabad to see the project through, led by N. Pant. Some had experience of microwave stations, but they had to start from scratch, searching the existing literature for clues and ideas. The United Nations Development Programme and the International Telecommunications Union provided expertise and equipment. Transmitters and receivers were provided by Nippon Electric in Japan.

FIRST EARTH STATIONS

In June 1967, the Experimental Satellite Communication Earth Station (ESCES) was duly opened at Jodhpur Tekhra in Ahmedabad to train Indian engineers and (others) in the value and use of satellite communication. The station, including its 14-m dish, was built in 87 days, and as soon as it was working, made contact with the American Applications Technology Satellite 2 (ATS-2) communication satellite. The satellite was tumbling out of control and the signal was not maintained but the breakthrough had been made.

Indian engineers next built the first operational Earth station in India, at Arvi near Poona. This had a dish more than twice the size, 29m, designed for overseas satellite communication via the international INTELSAT system. Consideration was given to buying in a system from abroad, but even though it took longer, it was decided to build all the equipment at home. A third station was then opened in New Delhi.

> **India's first Earth stations**
> 1967   Ahmedabad
> 1972   Arvi
> 1975   New Delhi

DEPARTMENT OF SPACE, INDIAN SPACE RESEARCH ORGANIZATION

Even as India carried out these landmark developments, steps were put in train to establish the space programme on a firmer, formal basis. The Indian space programme was formally organized with the setting up, by the government, of the Indian Space Research Organization, ISRO, established on 15th August 1969, the operating body, and later, in June 1972, of the Space Commission, the main policy-making body. The Space Commission operated under the Department of Space, DOS, under the office of the Prime Minister.

> **Directors of ISRO**
> 1972   Professor M.K.K. Menon
> 1972–84   Professor S. Dhawan
> 1984–94   Professor U.R. Rao
> 1994–   Dr Krishnaswamy Kasturirangan

BEGINNING OF APPLICATIONS WORK

Indian scientists were impressed from an early stage with the ability of satellites to map the Earth from orbit. Indian scientists, like Vikram Sarabhai and Professor P.R. Pisharoty (later called 'the father of remote sensing in India') quickly realized the immense potential of remote sensing in India. The country lost each year Rs50bn (€1bn) to pests, Rs15bn (€319m) to droughts and Rs7.7bn (€163bn) to floods—but satellites could give warning of all these dangers. One of the world's worst storms took place in November 1970 when a 9 m high sea surge headed into the Bay of Bengal at over 224 km/hr over half a million people were drowned, mainly in Bangladesh, but there were no satellite warnings then. 75% of India's population lived off agriculture and 60% lived in rural villages—and satellites offered a quicker means of reaching and warning them than any other method. And although this was poorly appreciated in the West, baseline Indian economic data were poor: maps in Asia were much less accurate than in Europe and few had been updated since the war. Again, satellites could rectify these gaps.

In 1970, scientists used a helicopter to detect coconut blight. In 1974, as part of the ARISE project (Agricultural Resource Inventory Survey Experiment), aircraft were used to survey agricultural fields in Andhra Pradesh and Punjab. Not long afterwards, India began to use *Landsat* data on a systematic basis. In 1974, the United States had orbited *Landsat*, the first of seven pioneering satellites devoted to such Earth resources studies.

In 1978, India began to build a station designed to receive data directly from American *Landsat* Earth resources satellites. Under an American–Indian agreement, a station was built at Hyderabad in Andhra Pradesh. This became the base of the Indian National Remote Sensing Agency, which analysed foreign and domestic remote sensing data.

Sadly, Vikram Sarabhai did not see the course which he charted for the Indian space programme come to fruition. He died suddenly of a heart attack on 31st December 1971 while visiting the Thumba launch range, aged only 52. Later, his stunned colleagues spoke of him much as Soviet space scientists spoke of the organizer of the Soviet break-through into space—Sergei Korolev. They spoke of his child-like enthusiasm for space travel, his vision of space applications in the service of Indian economic development, his ability to organize, lead, manage and inspire. At a personal level, they spoke of his

Early Earth station development

consideration, his humanity, his warmth, patience, his trust of people to do the job assigned to them. They admired his interest in art, music, literature and politics which matched his passion for science, technology and social reform. He always urged scientists to be involved in their countries and implored them to put science to work in the service of practical needs. He himself endlessly proposed projects for communication, agriculture, family planning and education. He was showered with awards during his lifetime and many more posthumously.

ATS: VILLAGE TELEVISION

A formative experience for the Indian space programme was ATS-6, or the Applications Technology Satellite 6. ATS was an experimental American satellite launched in May 1974 with an unusually large 9.1-m antenna for direct broadcasting and was used for pioneering such diverse activities as teleconferencing and broadcasting medical advice. In what was called the Satellite Instructional Television Experiment, SITE, NASA moved ATS over India for an experiment in direct educational broadcasting for the period August 1975 to July 1976. One thousand two hundred hours of ATS programmes, made by the Space Applications Centre in Ahmedabad, were relayed to 2400 villages and were seen by an estimated five million people. The signals were relayed up from Ahmedabad and New Delhi to ATS and then retransmitted for four hours every day. The programmes covered such subjects as agriculture, health, hygiene, family planning, rural development and literacy. The 3 m diameter ground sets were made out of inexpensive equipment such as chickenwire and signal converters.

In a follow-up experiment called STEP, or Satellite Telecommunications Experimental Project, tests of ground communications were made from 1977–79 with Europe's satellite *Symphonie*. This concentrated on the testing of truck and jeep-based sending and receiving stations, the development of multilingual educational satellite television, the use of satellite communications in response to natural disasters and for broadcasting other matters of national importance (viewers noted that cricket matches fell within this definition).

By the time the ATS-6 experiment had concluded, India was ready to go for the historic step of developing its own Earth satellite.

PREPARATIONS FOR FIRST SATELLITE, *ARYABHATA*

India had signed its first space agreement with the Soviet Union in 1962. This was a low-key arrangement, principally giving the USSR access to the Thumba range for meteorological sounding rockets. The first major space agreement between India and the Soviet Union was signed in summer 1972. This provided for the launch by the USSR of an Indian satellite and for Soviet use of Indian ports by tracking ships and vessels launching sounding rockets. These negotiations and the satellite project were directed by Professor U.R. Rao. He was given, with his colleagues, 36 months to conceive, design and build a test satellite. 'We worked night and day in asbestos-roofed sheds with practically no infrastructure', he recalled later.

Aryabhata, India's first satellite, weighed 360 kg, and was launched by the Soviet Union on 19th April 1975 into an orbit of 563–619 km and dedicated to the study of stellar X-rays, neutron and gamma radiation from solar flares and particles and radiation fluxes in the Earth's ionosphere. *Aryabhata* was named after the fifth century Indian astronomer and mathematician. The Soviet Union used its *Cosmos 3M* rocket from the family of launchers which had been putting scientific and military satellites into orbit since 1964. Sadly, the power supply broke down after 60 orbits when an Indian transformer failed and the mission was abandoned after five days, although Soviet media reports long insisted that the mission was going perfectly and returning vast quantities of data [11]. As a result, a standby transformer was installed on the next satellite. The cost of the mission was Rs40 m (€800 000).

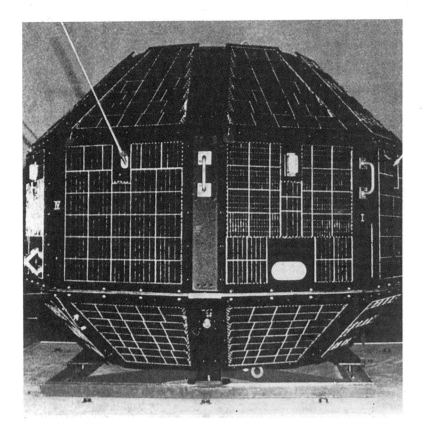

India's early Earth satellites were launched by the Soviet Union

Aryabhata eventually burned up in Earth's atmosphere on 19th February 1992, on its 92 875th orbit. A ground station at Sriharikota picked up the last 30 sec of transmission before it fell silent forever.

BHASKHARA

India's second satellite, *Bhaskhara*, launched on a *Cosmos 3M* rocket from Kapustin Yar near the banks of the river Volga, was in reality the back-up model of *Aryabhata*. *Bhaskhara* was named after a seventh century Indian astronomer. It was a 26-sided polyhedron, 1.19 m tall, 1.55 m in diameter, weighing 442 kg, with a surface area of 6.5 m^2. Energy was supplied by the 26 solar panels that covered the outside of the spacecraft, these cells feeding nickel—cadmium storage batteries. The spacecraft was controlled by an infrared horizon sensor, a sun sensor and triaxial magnetometers. The USSR supplied 3500 solar cells and the nitrogen pressurization system.

The satellite had four experiments. *Bhaskhara* carried two low-resolution television cameras, designed to show to best advantage changes in water and vegetation. Each picture was to show an area of 341 km^2 with a resolution of 1 km, providing black-and-white pictures from which false colour could be derived using digital processing. *Bhaskhara* carried three radiometers to study the oceans around India, obtaining information on sea surface temperature, ocean wind velocity and moisture content. The third experiment was a data collection and relay package, designed to pick up data from eight meteorological stations and retransmit them to a central receiver station in Ahmedabad. The fourth experiment was a pinhole camera to identify X-rays.

Bhaskhara entered an orbit of 519–541 km, 50.67°, 95.16 mins on 7th June 1979. Although the radiometers worked properly, the television cameras were not turned on for nearly a year because of gas being trapped in the camera system. They were used only during passes over India and were switched off otherwise in order to save power. About ten pictures were received every day. The cameras provided information on snow melting in the Himalayas, river flooding in northern India, desertification in Rajasthan, rainfall off the coast of India and mineral resources in Gujarat. Cost of the project was Rs65m (€1.38m).

SUCCESS OF *BHASKHARA 2*

Bhaskhara 2 was launched by the Soviet Union in November 1981. Its cameras provided information on agriculture, weather, and vegetation. Its images were used in preparing land-use maps in west Bengal and the microwave sensor contributed to knowledge of sea winds, water vapour and rainfall rates. It seems to have been much the most successful of the first three launches and returned the most data. *Bhaskhara 2* burned up on 30th November 1991.

The early launches			
Aryabhata	19 Apr 1975	Kapustin Yar	*Cosmos 3M*
Bhaskhara 1	7 Jun 1979	Kapustin Yar	*Cosmos 3M*
Bhaskhara 2	20 Nov 1981	Kapustin Yar	*Cosmos 3M*

Bhaskhara 2

The *Aryabhata* and *Bhaskhara* satellites were constructed by Hindustan Aeronautics Ltd (HAL), only the solar panels, batteries and thermal paints being supplied by the Soviet Union. The launches were provided free, on the terms that the USSR had access to the data collected.

AN INDIGENOUS INDIAN ROCKET

Ultimately, India did not wish to depend for long on other countries, however well-meaning, to launch their satellites (the USSR) or develop satellite programmes (the USA). Even as the *Aryabhata* launch was planned and the ATS experiment took place, efforts were under way to develop a home-built rocket launcher. There was a rapid expansion in the space industry in the mid-1970s. The numbers employed rose to 10 000, many of whom were professional scientists persuaded to return from abroad. By this stage, 30 universities were involved in space research projects stimulated by ISRO.

The development of an indigenous launcher, the Satellite Launch Vehicle (SLV), began in 1973, with the objective of placing a small, 40-kg satellite into an orbit of 300 to 900 km. 46 public and private enterprises were involved in the venture. The SLV had 10 000 components, of which 85% were developed in India itself. The SLV was a four-stage rocket, 22.7 m high and 1 m in diameter, making it unusually slim. The SLV was by world standards, tiny in size, about halfway between the Japanese *Lambda* and *Mu-3*. In effect, India hoped to build a rocket comparable to the American *Scout*, which had been placing small payloads in orbit since the early 1960s (although superior American engineering techniques gave *Scout* a payload of 180 kg). The fuel was made at the launch site of Sriharikota and comprised 70% ammonium perchlorate and 30% aluminium. Years later, the Americans chose similar fuels for the solid rocket boosters of the space shuttle.

The SLV had a take-off weight of 17 tonnes. The first stage was 10 m long, with one 63 500-kg thrust motor and the solid fuels were packed in three segments. The second stage was 6.4 m long, with a single motor and the third stage only 2.34 m long. Finally, the 1.5 m long fourth stage had a motor of just over two tonnes' thrust.

SLV launch vehicle		
Length: 23 m Diameter: 1 m Weight: 16.9 tonnes Performance: 40 kg to 300–900 km orbit		
Stage	Type	Thrust, kg
1	Solid	63 500
2	Solid	27 200
3	Solid	9 200
4	Solid	2 700

PLANNING THE FIRST HOME-LAUNCHED SATELLITE

The first domestically launched satellites were designed as technology demonstrators. They were called *Rohini*. The first series comprised 35-kg multi-sided capsules with a charge-coupled device (CCD) camera with 1-km resolution for Earth observation. A key objective was to test the Indian-made solar cells. Originally, India had planned its own first satellite for 1974, but there were long delays in the supply of key components and ISRO engineers had to be seconded to the companies concerned to speed things up. They also missed the deadline of 1977, the thirtieth anniversary of Indian independence.

The establishment of a domestic rocket and satellite programme required the building of a rocket launching base. Hitherto, launches had been made from Thumba and another site, Belasore, was also to be used for sounding rockets. For the national rocket programme, a new site was chosen, Sriharikota Island on the southeast coast.

Sriharikota has an area of 180 km^2, a barrier island sandwiched between the Bay of Bengal and Pulicat Lake. The island is 44 km long and 7.8 km broad at its widest, and has two streams in the middle, the Peddavagu and the Chinnavagu. The island is a sand ridge, at no stage more than 10 m above sea level. The site is not unlike Cape Canaveral, which is also a sand ridge with waterways. Annual rainfall is 1200 mm, moderate from July to

India's launch sites: Sriharikota and Trivandrum

September but very heavy during October to December. Like Cape Canaveral, Sriharikota abounds with wildlife, with 400 species belong to 110 families. Some are unique to the island. A herbarium was established to preserve the flora and fauna there in 1989 with scientists from Sri Venkateswara University.

FIRST LAUNCHES: INDIA—A SPACEFARING NATION

In the first attempt to orbit a 36.7 kg *Rohini* satellite on 10th August 1979, the vehicle suffered a malfunction and crashed into the Bay of Bengal 317 sec after launch, 500 km downrange. Nitric acid had drained off into the guidance system, it was later discovered. Space engineers formally declared the mission to be '70% successful'.

The first successful *Rohini* was eventually launched on 18th July 1980. The 0.5 m diameter, 40-kg spheroid satellite entered orbit of 306–919 km, 44.75°, 96.85 min. The solar cells managed to develop 3 W of electrical power and the satellite spun at the rate of 165 rpm. The satellite was tracked by ISRO stations in Sriharikota, Nicobar, Thima and Ahmedabad. India had become the seventh space nation.

The second *Rohini*, also a 40-kg spheroid, reached an orbit of 187–418 km, 46.27°, 90.49 min on the SLV from Sriharikota on 31st May 1981. It was hoped that its imaging camera would provide quality Earth resources pictures. The perigee was much lower than planned and the satellite burned up on 9th June after only 130 revolutions. Something appears to have gone wrong at the start of the firing of the fourth stage.

The third satellite was much the most productive. The red-and-white four-stage rocket, with the Indian national flag on its side, took off watched by Prime Minister Indira Gandhi. Afterwards, she congratulated the 800 personnel of the launch team and told them they had done a 'great job'. At a subsequent press conference, she justified Indian space spending on the basis that it would solve agricultural problems. Like a child's education, she said, the results come later. *Rohini 3* entered orbit on 17th April 1983 from Srihasikota. An hour and a half later, it was tracked for 11 min on its first pass over India. The satellite had a camera capable of picking out water, vegetation, clouds and snow. Weighing 41.5 kg, its two cameras sent back over 5000 images before the satellite was turned off 18 months later.

India received welcome recognition from the international astronautical community when in 1988 the 39th World Congress of the International Astronautical Federation was held in Bangalore. Six hundred technical papers were presented at the event which was attended by almost nine hundred delegates from countries all over the world. The hosts were India's own astronautical society, located in Bangalore.

CONCLUSIONS: A DISTINCT PATH OF DEVELOPMENT

The Indian space programme grew out of the sounding rocket programmes of the early 1960s, the building of Earth stations in the late 1960s and the development of Earth resources programmes in the early 1970s. A distinctive role for an Indian space programme was mapped out by its founder, Vikram Sarabhai, one whereby the programme would concentrate on Earth resources, applications and telecommunications. Although India often bought in the early stages of development from abroad—using America's ATS to demonstrate educational television and the Soviet Union to launch its first three satellites—a constant theme was the priority given to indigenous design and construction. By the early 1980s, India had joined the world space launcher club, the first developing nation to do so.

9

Space technology for development

By the early 1980s, India had become a spacefaring nation. The country had launched three satellites of its own and had three sputniks orbited by the USSR. The experimental phase was over: now was the time to put space technology to the operational service of Indian economic and rural development, the purpose for which the enterprise had originally been intended. Two main systems were developed: the Indian Remote Sensing Satellite System (IRS) and the Indian Satellite System (INSAT).

INTRODUCING IRS

The next stage of development was the IRS or Indian Remote Sensing Satellite. IRS was part of the Indian Remote Sensing Satellite System, itself in turn part of the National Natural Resources Management System.

IRS was designed to weigh up to a tonne, cross the same stretch of Earth every 21 days and carry out systematic surveys of the Earth's surface. IRS was designed to carry three linear imaging self-scanning sensors, one 70 mm, two 35 mm, for the use of agriculturists, geologists and hydrologists.

IRS-1A, THE MAPPER

The first IRS, IRS-1A, was launched by the Soviet Union in March 1988 on a *Vostok-Meteor* rocket and operated successfully for three years. The 940-kg satellite was put into a 1112-km polar orbit of 103.2 min. Fifty Indian scientists, accompanied by the press, travelled to Baikonour to supervise the launch. Unfortunately they saw little of the take-off, which was obscured by fog and snowstorms. India is believed to have paid about €2m for the launch and the satellite cost Rs650n (€13.8m) to build.

On its third anniversary, a postage stamp was issued in its honour. By then, IRS-1A had made 50 complete maps of India and taken part in a number of remote-sensing experiments. The all-India maps provided disturbing information that the national level of forest had decreased from 14% to 11%. On the positive side, water maps had led to an increase in the success rate of bore-drilling from 45% to 90%. A full national inventory of waste land had been made on a 1:250 000 scale, one which suggested that at least half could be reclaimed. Crop yields had been estimated on the basis of its data.

By its fifth anniversary in 1993, it had completed a salt map of the country on a 1:250 000 scale and it had returned a total 400 000 images in 25 470 orbits, of which 3700 were passes to map India. It was eventually retired in 1995 and put in reserve.

IRS-1B, 1C: THE EARTH, IN UNPRECEDENTED DETAIL

The second, IRS-1B, weighing 908 kg was put into orbit by *Vostok* rocket from Baikonour cosmodrome into a 857–918 km polar orbit on 29th August 1991. This time the price charged by the now cost-conscious Russians had risen to Rs629m (€13.39m). It had a bank of linear imaging scanning cameras and sent back its first pictures the next day. It was specifically designed to focus on forecasting the crop yields of tea and coffee. Three years later, IRS-1B had covered India fifty times on cycles of 22-day passes.

The third, IRS-1C, weighing 1250 kg, was launched from Baikonour on 28th December 1995 using a *Molniya M* rocket, putting the satellite in a 816–818 km polar orbit. The rocket flew an unusual 901 sec profile, the fourth stage being used for orbital injection. IRS-1C was an advance on 1A and 1B, offering improved resolution, stereo viewing and frequent site revisits, making it the most advanced remote sensing satellite in the world at the time. Solar panels sprang out 93 sec after orbital injection and the spacecraft swivelled to acquire Sun lock so as to start generating electrical current at once. IRS-1C carried a tape recorder so that data could be stored for later transmission.

> **Tasks of IRS-1C**
> Crop inventories
> Detection of crop pests
> Mapping of forests and tree types
> Town and city mapping
> Detection of changes in land use
> Soil mapping, detection of erosion
> Inventory of water resources
> Environmental monitoring
> Detection of silting
> Flood damage assessment
> Geological mapping

Within ten days, its 10-m high-resolution camera was switched on and the satellite declared operational after a month. IRS-1C carried a panchromatic camera, a linear imaging self-scanning sensor operating in four bands and a wide-field sensor with a swath of 810 km and a resolution of 188 m. Data were transmitted in the first instance to Hyderabad but also to the United States, Korea, Japan, Germany and Dubai. Four years later it had completed its primary mission and still had sufficient fuel to go on. Some of the detail available on the satellite images was extraordinary: the camera had sufficient resolution to pick out passenger aeroplanes parked at airports.

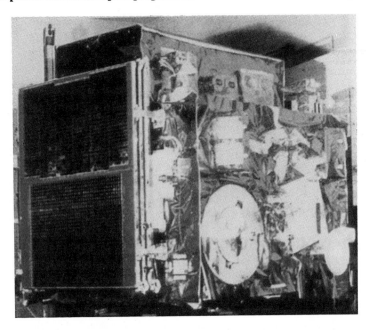

IRS-1A

IRS marked the introduction of the Linear Imaging Self-Scanning Sensor, or LISS, which operated in the visible and infrared wavebands using French-made charge-coupled device scanners. IRS-C introduced a panchromatic camera able to provide 6-m resolution from 600 km, then thought to be the highest available in the world.

In 1993, India gave a franchise to EOSAT, an American company based in Lanham, Maryland, to process data covering countries other than India and to sell images to the world market (it already held the rights on American *Landat* data). From June 1994, EOSAT began to receive the IRS data at its station in Norman, Oklahoma. Another indication of the quality of IRS data was that in 1998, the Japanese National Space Development Agency, NASDA, a world leader in the area, applied to receive IRS data at its own Remote Sensing Technology Centre.

INDIAN REMOTE SENSING—A BALANCE SHEET

The use of remote sensing by satellites in India is part of the national system for managing natural resources in India. This is called the National Natural Resources Management System (NNRMS), which has the brief of compiling satellite and ground data for purposes of combating drought, mapping waste land, estimating crop yield, water resources management, flood prevention, mineral development, land-use mapping, forestry management and ocean development (including coastal development and fisheries). The Department of Space has overall responsibility for the NNRMS. Local development plans for sustainable development have been drawn up for 45% of the country using IRS data.

First stages of the IRS system		
1A	15 Mar 1988	*Vostok*
1B	29 Aug 1991	*Vostok*
1C	28 Dec 1995	*Molniya M*

Among the benefits of remote sensing from satellites, the following have been cited:

- Satellite maps have tracked sewage entering rivers (e.g. the Yamuna, near Delhi) and have planned the location of treatment plants accordingly
- Studies of forest cover have mapped the space available to tigers
- Satellite pictures have showed the devastating effects of deforestation and mining
- Chemical, mining and engineering companies were asked to use satellite pictures before locating new sites, in order to prevent pollution of ground or surface water
- Ocean maps have indicated where fish are likely to be found, doubling catches
- Satellite photos have identified poor embankments on rivers, where breaches are most likely in the event of future flooding
- Soil maps have been compiled showing the respective levels of salt and lime
- Remote-sensing satellites have been important in mineral development. IRS-1A located deposits of diamonds and rubies near Najranagar, zinc in Wantimata and copper in Karnataka. Other IRS satellites helped to find base metal, tin, iron ore, bauxite and oil.

The availability of water is one of the great problems of India, as it is in most developing countries. Under the National Drinking Water Mission, set up by the government in 1987, there is a commitment to ensuring every village access up to 40 litres-a-head of clean, drinking water a day. The Department of Space was charged with making space-based maps to identify the presence of underground water. Accordingly, water maps of 447 districts on a 1:250 000 scale were compiled. These are coloured maps, with blue for areas of good groundwater potential, green for areas of moderate potential depending on the geology, red for no potential and yellow for doubtful. At the other extreme, IRS satellites have tracked flooding, enabling warnings to be provided and safer areas to be predicted.

The biennial national forest survey of India is now satellite-based. The 1993–95 survey, for example, showed a marginal increase in the total forest cover in the country. Related to this is wasteland mapping. Here, satellites carried out a national mapping exercise for the National Wasteland Development Board at 1:250 000 scale, enabling many areas to be reclaimed for food or forestry. One of the early achievements of the IRS system was the compiling of 1:250 000 maps on agricultural use, able to distinguish cropping patterns, grazing land, waste areas and waterlogged parts.

The 60 km long ring road for Bangalore was planned using satellite data. This approach had the advantage of being more up-to-date than conventional maps (which had not included new housing in the path of the proposed road), enabled the road to follow the geological terrain, was faster than ground-based surveying and realigned the road on poor quality land.

Satellite imaging was used in an experiment to promote the silk industry in India. Despite great demand for Indian silk, production is limited to a small number of areas. IRS-1A was used to estimate crop yields in four districts in Bangalore, Mysore, Manya and Kolar, to identify new sites for mulberry growth, and to forecast silk cocoon production.

> **Crops assessed for yields by IRS**
> Wheat
> Rice
> Mustard
> Cotton
> Groundnut
> Bajra
> Sugarcane

In an experiment off the three maritime states of Gujarat, Maharashtra and Andhra Pradesh, IRS data of sea conditions were used to predict the location of fish. Generally, fish may be found in areas where temperatures change and not in areas of uniform temperature. Local fishermen headed for the identified areas, with dramatic results in increased catches.

In an unusual exercise to protect wildlife, the IRS system was used to map elephant trails in north Bengal so as to better assess the suitability of their habitats. In Junagadh, satellite imagery found an unexplored archaeological site which subsequently yielded pottery, bones and stone walls.

IRS data has been used to warn of locusts, especially the desert locust which thrives in Rajasthan, Haryana and Gujarat. A particular form of stratus cloud is a known carrier of wheat rust spores, so tracking these clouds by satellite can give 20 to 25 days warning of the arrival of wheat rust, thereby enabling precautions to be taken. Interpretation of IRS data was able to identify conditions favourable to the spread of brown planthopper, a nasty pest which can ruin rice production. Similarly, areas infested by cotton fly could be identified on satellite photos (they show up as dark red on a lighter red background)—the only remedy being crop rotation to break the life cycle of the pest.

IRS pictures used a series of false colours derived from the different scanners to throw sharp relief on the ground below. Early IRS pictures clearly showed flooded areas, different kinds of grasses, water holes for animals, different types of soils, vegetation types, soil erosion, forest fires, sediments off coasts, damage done by mining and human settlements.

By 1997, the IRS system was used systematically for assessing likely national crop yields for seven main crops and was used to test for others (e.g. chilli).

In the course of 1994–95, the parliamentary standing committee on science, technology, environment and forests made a detailed review of the work of the Department of Space in remote-sensing applications and satellite-based early warning. In a lengthy report examining all aspects of the programme, the parliamentary committee concluded that the programme was 'well thought-out and implemented on a steady long-term basis so as to derive optimal advance [for India] from this high-tech area'. It was a unique programme in the world and commended the efforts of the Department of Space.

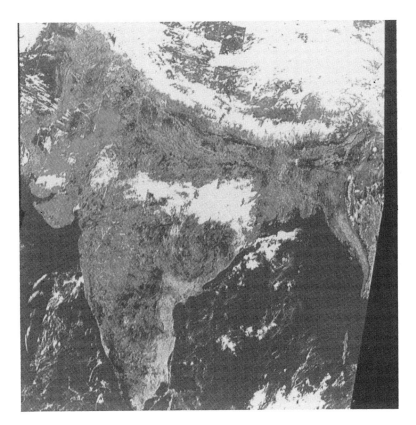

India from space—from IRS

INSAT: INDIA'S COMMUNICATION AND WEATHER SYSTEM

The IRS system is one leg of the Indian applications programme: the other is INSAT.

The INSAT (Indian National Satellite System) programme was planned in 1976 as a Rs19bn (€405m) project, a joint venture between the Department of Space, the Indian Meteorological Department, the Department of Telecommunications and All-India Radio. The aim was to provide a communications and weather satellite system for India. A head office for the system was set up in Bangalore.

Such satellites must be sent into 24-hr synchronous geostationary orbit, something far beyond India's capabilities. Accordingly, foreign launchers had to be used. Some time was spent in these negotiations. An agreement was signed in July 1978 between the space commission and NASA whereby the United States would launch the first Indian National Satellite System satellite, INSAT 1, on the space shuttle in 1981. Nor did India have the know-how to build such a system itself, so the first round of INSAT satellites, called the INSAT 1 series, was also sent out to tender abroad. The satellite would be built by Ford Aerospace and was the first agreement whereby NASA would launch a commercial satellite on the shuttle for another country. Two would be built and flown, INSAT 1A and 1B.

With the INSAT series, India attempted to combine the two great advantages of geosynchronoous orbit: communications and a high-altitude platform from which to observe the Earth. Most other countries have used 24-hr satellites for one purpose or the other, but not both at the same time. India was the first country to combine these two functions, an approach which it reckoned was 40% cheaper in the long run than using separate satellites.

At that time, both television and telephone services in India were poorly developed. Even by the late 1970s, television reached less than 30% of the people. Although television had started in India in 1959, there had been little investment in the system and a second station had opened only in 1972 in Bombay. Those programmes which were not imported were of poor quality and in black and white, contrasting unfavourably with India's robust film industry. Television sets were in short supply, a function of restrictions on imports and the lack of domestic production. The ground station system was poorly developed, and India had to build 30 Earth stations in quick order, achieved through importing American and Japanese equipment at a cost of Rs2.664bn (€56.7m). The aim of the INSAT series was ambitious: to more than double the proportion of Indians receiving television to 75% of the population.

India set up ground stations to handle INSAT 1 in Hassan, New Delhi and Madras and a number of truck and jeep-mounted terminals for use during natural disasters like floods. Hassan had two 14-m dishes. Weather pictures were to be sent down to the Earth station in Delhi for passing on to the weather ministry for further relaying to secondary stations.

INSAT PRECURSOR: APPLE

Fortuitously, an opportunity arose to test out some of the principles of the INSAT system before the first shuttle mission was due to fly. Europe was then in the early phases of testing its new commercial launcher, the *Ariane*.

The first four flights of *Ariane* were development flights. The European Space Agency (ESA) offered free space on these missions, on the understanding that, being test flights, there was a risk of failure. India already had good relationships with ESA, having signed its first agreement with the European Space Agency in 1971, followed by further major agreements in 1977 and 1978. India responded promptly to the ESA opportunity, developing an experimental communications satellite on a tight time schedule. The satellite was called the Ariane Passenger Payload Experiment (APPLE). Design began in 1978, fabrication in early 1979 and the flight model was completed in 1980. The cost of the project was Rs150m (€22.9m), the spacecraft being built by Hindustan Aeronautics.

The purpose of APPLE was to test out means of stabilizing satellites in 24-hr orbit, C-band communications transponders, the kick motor, solar panel and batteries. It was a major opportunity to develop the experience necessary for controlling satellites in 24-hr

APPLE paved the way for the INSAT system

orbit, and later in constructing a domestic communications satellite. Already, Indian space planners were thinking ahead to a domestic-built comsat, the INSAT 2.

APPLE was a 630-kg cylindrical structure, 1.2 m tall and 1.2 m in diameter, with two 1.2-m^2 solar panels and two transponders. The fourth stage of the SLV rocket was used as an apogee kick motor to get APPLE into its final orbit.

APPLE was launched by *Ariane* V3 (V = vol, or 'flight' in French) on 19th June 1981, accompanying the main payload, the *Meteosat 2* weather satellite. Despite a solar panel which jammed, APPLE relayed television programmes and educational teleconferences. The APPLE experiment lasted just over two years. APPLE involved the testing out of small portable terminals as little as 1 m across in states such as Orissa and Gujarat.

FIRST INSAT 1: A SYSTEM ESTABLISHED

The shuttle suffered many delays before it entered service and INSAT-1A was lofted into orbit instead by a *Delta 3910* on 10th April 1982 from Cape Canaveral. Owing to the lack of a suitable orbital slot, it had to be parked over Indonesia, which was far from ideal. INSAT 1 was built with 12 television transponders, two TV direct broadcasting antennas and a Very High Resolution Radiometer to image the Earth every 30 min. The radiometer had a resolution of 2.5 km in the visible band and 10 km in infrared. The TV transponders could reach up to 100 000 small Earth terminals. Instead of transmitting television, the system could handle up to 8000 telephone calls at a time or be used for radio. INSAT 1A was a box-shaped body weighing 1152 kg (fuelled, about 550 kg unfuelled) with a single solar panel with 12 942 solar cells at one end, counterbalanced by a boom at the other, the first such arrangement at the time. INSAT had a 445 N booster rocket to reach apogee and six 22 N thrusters for station-keeping.

Initially, the main antenna failed to fully deploy, threatening the broadcasting aspect of the Rs5.7bn (€122m) mission, but these problems were overcome, which freed the system by bathing the antenna in sunlight and by the brief firing of a thruster. The meteorological system was designed to provide half-hourly pictures of cyclones and sea conditions around India, warn of floods and disasters, and collect and transmit data from remote observation stations. The first pictures were received in early May 1982 with a night-time image and then a daytime one. The satellite worked well initially and produced good pictures, but during the late summer the attitude control system depleted its reserves, station-keeping proved impossible and the satellite was taken out of commission on 4th September after being operational for only four months. The timing was especially unfortunate, for it coincided with the opening of the Asian Games, hosted by India, and the country had to buy in communication satellite lines from the United States and Soviet Union to make good the loss.

INSAT 1B was launched from the payload bay of the space shuttle *Challenger* on the eighth shuttle mission, STS-8, on 30th August 1983. The astronauts spun the blue and gold INSAT out of the shuttle's payload bay at 40 rpm. The astronaut responsible, Guion Bluford (also the first black in space), reported that 'INSAT was deployed on time with no anomalies and the satellite looked good'. 45 min later, the payload assist module fired to send INSAT into a geostationary transfer orbit of 296 km by 38 173 km. Following several firings by the satellite's own liquid-propelled engine, INSAT was stabilized at 74°E (though there were difficulties with extending the solar array and this was not accomplished for over a month). INSAT 1B extended television coverage to 70% of India's population. In 1991, experiments were made in videoconferencing to train adult education officials. ISRO is reported to have paid NASA Rs1.485bn (€31.6m) for the launch. No one was apparently more relieved than prime minister Indira Gandhi, who was anxious to enlist television in her campaign to be reelected the following year. 1B became an early star of the programme and soldiered on until 1993 when it was finally retired.

Although it had been planned to fly INSAT 1D on the shuttle in 1990, the new post-*Challenger* disaster shuttle programme moved away from commercial satellite launches. INSAT 1D was transferred to a *Delta*. In a bizarre accident on the launch pad on 19th June 1989, a crane hook tangled with cables and crashed into the satellite, destroying the C-band antenna and damaging the rest of the satellite so badly that there were fears it would be written off. The manufacturers, Ford, deemed the satellite repairable and took it away for a year's repairs which cost as much as €9.4m (Rs443m). However, Ford brought it to its workshop in Palo Alto, California, just in time for the big earthquake there, where it suffered a further €140 000 (Rs6.5m) damage. INSAT 1D was eventually launched on *Delta 4925* on 12th June 1990. The four INSATs had used, between them, three different launchers (*Delta, Ariane* and shuttle). INSAT 1D was much the most successful of the

Second generation: INSAT 2C

four and was still operational five years later, having sent back 24 500 weather images. Television had been relayed to 2000 community receiving sets.

INSAT 2: MADE AT HOME

The Rs20bn (€425m) INSAT 2 programme was approved in April 1985, the principal aim being to develop satellites of INSAT 1 standard within India itself, although foreign launchers would still be required. Ultimately, the intention was that later versions would be fired by Indian launchers to geosynchronous launchers. The new satellites were 50% heavier than INSAT 1, carried 18 transponders and a beacon for an international maritime distress system. Each satellite weighed 1906 kg at launch, 911 kg once all its propellant had been depleted. The solar wing had a span of 23 m and provided 1400 W of power. Station was to be maintained by 16 attitude thrusters. The Very High Resolution Radiometer had improved definition compared to INSAT 1 and was able to send a full Earth scan every 33 min or examine more selected areas for cyclones every 7 min.

Europe's *Ariane* was chosen as the launcher for the series. The INSAT satellite was built in India itself, with some limited assistance from abroad. Britain was involved in the INSAT 2 design, British Aerospace supplying titanium-made helium pressurization tanks. The electronic modules and high-speed, high-density integrated circuits were designed by Racal-Redac, based in Reading. The aim of INSAT 2 was to provide business communications not only throughout India but to a larger telecommunications footprint taking in southeast Asia and the Middle East. Annual revenues quickly reached the order Rs5bn (€85m) a year.

The rocketry used to put the INSAT in its final orbit was developed by the Liquid Propulsion Centre in India. Called the Liquid Apogee Module, or LAN, its function was to provide 6000 sec of thrust in one burn or more to get the communication satellite from an orbit of 200 by 36 000 km to a circular one of 36 000 km. It had a thrust of 440 N and a specific impulse of 310 sec, and used mixed oxides of nitrogen and monomethyl hydrazine.

Launched by *Ariane 4* on its 51st flight, V51, on 10th July 1992, INSAT 2A entered a geostationary transfer orbit 10 min later. The master control facility at Hassan took over control and manoeuvred it into position for the critical firing to raise its perigee and place it in geostationary orbit. The next day, the liquid apogee motor fired twice, once for 3900 sec and then for 806 sec. Three day slater, the 15.5-m^2 1200-W solar array was deployed, followed by the Sun tracker, the antennas and the boom. On 23rd July, Hassan commanded that the orbit be trimmed and that the satellite drift toward its parking slot at 74°E, which it reached on the 29th July. By this stage, the transponders and the weather imaging system had been switched on.

INSAT 2A began operations in August 1992 and the Minister for Science and Technology, Mr Kumaramangalam, was able to tell the Indian parliament, the Lok Sabha, that India could be justifiably proud of having developed such a state-of-the-art communications satellite by itself. With INSAT 2A, India was able to operate a series of telephone trunk routes across the subcontinent and to provide television for 65% of the Indian landmass and 80% of the population, and signals for 127 radio stations. In September, the weather imager on INSAT 2A sent back a picture of the Earth's globe: to the north were the dusty browns of Arabia; to the south, the swirling clouds of the southern ocean; and in the middle, India, with three large cloud areas off the coast.

INSAT 2B was launched by *Ariane 4* from Korou, French Guiana, on 23rd July 1993. On 6th August it beamed its first weather images back to Earth and it was declared operational shortly thereafter. It carried search-and-rescue transponders under the international COSPAS/SARSAT system. This is a global network whereby satellites can pick up a distress beacon, generally from ships at sea and relay the signal to the nearest ground station. Because it uses a global network of polar-orbiting satellites, most distress calls can be notified to a ground station and then to a national rescue service anywhere in the world within half an hour. India installed its first COSPAS/SARSAT ground station, called a local user terminal, in Bangalore in 1989 and a second in Lucknow the following year. India installed a 2.4 m dish, which not only served distress calls adjacent to India but acted as a relay service for Bangladesh, Indonesia, Kenya, Malaysia, Maldives, Singapore, Somalia, Sri Lanka, Tanzania and Thailand as well. The most publicized rescues under the system have been those of round-the-world yachts, but the unpublicized work of the system goes on all the time. As often as not, alarms are triggered by cargo ships either sinking or in danger of doing so. In 1998, a beacon was set off when a trekking expedition in the Himalayas got into trouble and had to be rescued.

INSAT 2C, with 23 transponders was launched by *Ariane* on 7th December 1995. It introduced Ku band technology to India and used such a wide footprint that the satellite could be used as far afield as Australia, China and central Asia. By then, India had 70 transponders in space.

INSAT 2D was launched by *Ariane* from Korou on 4th June 1997 but owing to problems with electrical supply it lost its ability to lock on the Earth and was lost. Communications to 83 terminals in the north and northeast of the country (Jammu, Kashmir and Uttar Pradesh) went down. Even the national stock exchange in Bombay was knocked out. The disrupted stations were transferred to the old INSAT 1B; the national stock exchange declared a holiday while the problems were sorted out. Ground control in Hassan managed to restore three, later seven, transponders on INSAT 2D and even reconnect the stock exchange, but this provided only temporary relief and the satellite was abandoned in October, a demoralizing outcome. India then bought the *Arabsat 1C* satellite from an Arab consortium which had operated the comsat since 1992 for €37.7m. (Rs1.77bn). *Arabsat* was moved to INSAT's orbital position where its 26 transponders could be brought into play by Hassan master control (it was posted in the name of INSAT 2DT). This was a temporary measure and the decision was then taken to bring forward the launch of the first INSAT 3.

INSAT 2E was launched by *Ariane* on 2nd April 1999 on its 117th mission. ISRO scientists in Korou cheered twice—at separation 21 min after liftoff and some hours later when the first signal was acquired. Their celebrations were short-lived, for the circulariz-ing of the orbit to 36 000 km went astray when the liquid apogee motor cut out 16 min into a 75-min burn. However, with three further firings commanded from Hassan, it eventually arrived on station. INSAT 2E was declared operational a month later. INSAT 2E weighed 2.55 tonnes, had a 14-m solar panel and boom, carried 17 C-band transpond-ers and a 1-m resolution CCD camera and, for the first time on INSAT, a radiometer for the assessing of water vapour. This radiometer had 2-m resolution in the visible land and 8-m in infrared. Such instruments have proved to be very accurate in forecasting the likely volume of rainfall. INSAT was also able to pick up and retransmit data from over a hundred remote location platforms in uninhabited areas. INSAT 2E was positioned 83°, over the Indian Ocean. The first weather pictures were received in Hassan Master Control Facility in mid-April. The satellite was expected to operate until 2011.

REACHING THE VILLAGES

By 2000, 700 television and almost 200 radio stations used space-based signals. 90% of the Indian population now received television, compared to 30% before the first INSAT

was launched. In the rural areas, 35 000 of India's 520 000 villages had satellite terminals. INSAT 2E's performance was such that the European INTELSAT consortium leased
11 transponders for 10 years for a cost of Rs8.86bn (€188m). The International Telecommunications Satellite Organization, Intelsat, which provided international lines to India,
noted that its Indian revenues rose from €16m (Rs752m) in 1996 to €29.2m (Rs1.372bn)
in 1998.

Typical Indian satellite dishes are in the 10 m diameter range, but the size is gradually
being brought down to 3 m and later 2 m. The later INSAT 2 series operationalized a
system first tested in 1987 whereby it could be commanded to trigger battery-power
systems in Indian villages to set off sirens to warn of impending cyclones. At one stage,
with a cyclone heading in, the alarms triggered the evacuation of 170 000 people and
many lives were saved. Typically, the siren klaxons loudly for a minute, followed by a
warning in the local language, giving warning of the impending danger and the
precautions to be observed. By 2000, 250 receivers were installed in the most cyclone-
prone areas.

India's satellite systems have been used for video teleconferencing, training for farm-
ers, programming for rural development workers, with networks set up under the Indira
Gandhi Open University and the National Dairy Development Board. In February 1995,
Prime Minister P.V. Narasima Rao dedicated a training and development channel on
INSAT. It was put to early use in the training of rural development workers. By 2000,
about 120 hr of educational television was going out each month to 4000 schools and
colleges. The Training and Development Channel was used to transmit distance education in rural development, women's concerns and child development. 375 terminals were
in use for this purpose, but with expansion set to reach the 2000-mark. It is an interactive
service, one which enables learners to phone back questions to the teachers (the system is
called *talk-back*). Later, digital television will enable lessons to be downloaded and
reused locally. By 2000, the INSAT system was providing 4700 voice circuits via 430
Earth stations and 1200 terminals. A unique experiment in development education
through satellite began in 1996. Television was supplied to the poorly developed region
of Padya Pradesh, evening educational television being provided in health care, literacy
and the management of water resources to 150 sites. It received such a positive response
that the programme was extended to a further 200 sites in the adjacent districts of Dhar
and Barwani.

PROMISE OF INSAT 3

Next in the series was INSAT 3, a series of five satellites ordered in 1998. Work on
defining the specifications for INSAT 3 began in 1992. The aim was to increase the
numbers of transponder available from 70 in 1998 to 130 by 2002 and later to introduce
digital television. INSAT 3 will have 16 transponders, including an education channel.
INSAT 3A will be launched by *Ariane* in 2000 at a cost of Rs3257m (€69.2m). INSAT
3A was a multipurpose communications, weather and search and rescue satellite with a
temperature sounder. 3B and 3C will be communications satellites only, while 3D will be
primarily a weather satellite with a six-channel imager and a 19-channel sounder.

INSAT series

APPLE	19 Jun 1981	*Ariane*
1A	10 Apr 1982	*Delta*
1B	30 Aug 1983	STS-8/*Challenger*
1C	21 Jul 1988	*Ariane*
1D	12 Jun 1990	*Delta*
2A	10 Jul 1992	*Ariane*
2B	23 Jul 1993	*Ariane*
2C	7 Dec 1995	*Ariane*
2D	4 Jun 1997	*Ariane*
2E	2 Apr 1999	*Ariane*
Scheduled		
3B	2000	
3A	2000	
3C	2001	
3D	2002	
3E	2003	

In a related project, India agreed to cooperate with France on a satellite project called *Megatropiques*, due in 2005, to examine tropical air currents and long-term climate patterns over the seas and continents.

Although India had no plans to be directly involved in the provision of low-Earth-orbit hand-held telephone communications, the ICO company chose Chattrapur as one of its 12 Earth stations, making an investment of €47.1m (Rs2.217bn) in the facility.

CONCLUSIONS

Thus by the late 1990s, India had established, through the use of foreign rockets, a comprehensive system of remote sensing, communications and weather forecasting. Even as it did so, there were further advances in domestic rocketry. The remote sensing system provided direct, tangible benefits to the Indian economy and was recognized as one of the most advanced in the world. The communications and weather warning systems not only provided a rapid modernization of Indian broadcasting and meteorology, but set a world standard for the imaginative use of the 24-hr orbit.

10

The new launchers

During the 1980s, India successfully flew an improved version of its first rocket and made major strides with the development of a new, powerful domestic launcher. More advanced Earth resources satellites were put in orbit. India joined the select group of countries invited to send guest astronauts into space.

ASLV

India's first launcher, the SLV, had very limited carrying capacity and no control systems. After the first successful launches of the Satellite Launch Vehicle, the SLV, India proceeded with the upgrading of the launcher in order to place larger, 150-kg payloads into 150–300 km orbit. The ASLV was essentially the *Scout*-class SLV, but with the addition of two strap-on boosters. The weight of the ASLV was almost 40 tonnes, the height 23.5 m. This time, the launcher was assembled vertically on the launch pad on a 40 m tall mobile service structure with lifts, access platforms and clean room. The strap-on boosters were first tested in flight in November 1985, when they were attached to a *Rohini RH-300* sounding rocket.

ASLV was seen as an intermediate rocket whilst a much larger one was in design, so only four ASLVs were commissioned. The extra thrust meant that satellites of up to 150 kg could be put in orbit with payloads of up to 35 kg. These were called 'stretched *Rohini*; because they were an extension on the earlier small *Rohini* satellites, being designed to carry scientific and sensing instruments. The new version of the *Rohini* satellite was larger than those put up by the SLV and typically carried gamma ray detectors, a monocular optical scanner, ionospheric monitor, two retro-reflectors for laser tracking and an X-ray instrument.

ASLV HEARTBREAKS

The ASLV failed on its first flight in March 1987, the second stage misfiring at 48 sec. An electrical short circuit was later blamed. The stretched *Rohini* (later called SROSS-A) satellite was lost. The second flight, on 13th July 1988, also failed when the flight control system malfunctioned and the rocket spun out of control. The top stage of the booster broke free at 50.4 sec at an altitude of 25 km, cartwheeled while still firing and impacted into the ocean after 257 sec aloft. The second stretched *Rohini* (SROSS-B) had eight solar wings, a monocular electro-optical stereo scanner and a gamma burst experiment.

A number of changes were made in the ASLV design after these two disappointments. Two fins were added in order to improve the stability of the launcher. The automatic control system was redesigned. Other procedures were set in place to reduce stress on the ·vehicle during its ascent.

Four years later, the ASLV finally succeeded in orbiting SROSS-C on 20th May 1992. The core stage and two strap-on rockets ignited at the zero-point in the countdown. The other four strap-ons fired at 30.5 sec, each running for 74 sec. First stage burnout took place at 111 sec, followed by staging. The second stage burned UDMH and nitrogen tetroxide for 150 sec before the third-stage solid rocket motor took over. This brought the vehicle to 421 km some 380 sec after liftoff. The rocket coasted for 3.5 min before the two small fourth-stage solid-fuelled engines ignited for 405 sec.

ASLV

Length: 23.5 m
Diameter: 3 m
Weight: 39 tonnes
Performance: 106 kg to 475 km circular orbit at 45°, 150 kg to low Earth orbit

Stage	Type	Thrust, kg
Strap-ons (2)	Solid	120 300
1	Solid	71 600
2	Solid	31 000
3	Solid	9 200
4	Solid	3 600

The final orbit insertion manoeuvre was not wholly successful, setting a much lower perigee than planned, though the satellite nevertheless met its mission requirements before the payload burned up after two months on 14th July. ISRO reported that it worked very well and they had successfully tested out the retarding potential analyser which scanned the Earth's equatorial ionosphere and troposphere for its thermal structure and electron densities.

SROSS-C carried a gamma burst experiment developed by ISRO in Bangalore. SROSS was cylindrically shaped with eight solar panels generating 45 W of power. The gamma burst detector was switched on during orbit 119 and after a few false alarms when it passed through the radiation belts recorded its first gamma ray burst on orbit 337.

Two years later, ASLV launched SROSS-C2 at night on 4th May 1994, the solid rocket boosters turning the dark sky to light as it headed skyward. The 113-kg satellite carried a gamma ray burst experiment and a retarding potential analyser to investigate the ionosphere and thermosphere at equatorial and low latitudes. It had a small motor with

---

Stretched *Rohini*—summary

1	24 Mar 1987	Sriharikota	ASLV (fail)
2	13 Jul 1988	Sriharikota	ASLV (fail)
C	20 May 1992	Sriharikota	ASLV
C2	4 May 1994	Sriharikota	ASLV

---

5 kg of fuel; in July, the satellite made a series of manoeuvres to lower the height of its orbit to reach its intended destination of from 429 km to 628 km, 46°, and still had 1.5 kg of fuel remaining. SROSS was octagonally shaped, carrying solar cells on its body and on attached solar panels. The detector found twelve gamma bursts in the first year in orbit and the analyser made 600 sets of ionospheric data. C2 was the first Indian satellite to have indigenously made batteries—twelve nickel–cadmium cells able to complete up to 18 000 charge–discharge cycles. It was still in operation in late 1999. This launch marked the conclusion of the ASLV series.

PSLV: INTO THE BIG LAUNCHER LEAGUE

The ASLV was always seen as an intermediate step toward India's second generation of launch vehicle, the Polar Satellite Launch Vehicle, the PSLV. The aim of the PSLV was to place 1000-kg remote-sensing satellites into 900 km sun-synchronous polar orbits. The PSLV was designed for the Indian Remote Sensing Satellite (IRS) series, which had thus far been launched by the USSR. In effect, the PSLV enabled India to join the 'big launcher' club of the US, Russia, Europe, China and Japan. The PSLV was almost ten times bigger and heavier than the ASLV, being 275 tonnes in weight and 44.1 m tall. It was broadly comparable to the American *Delta* and its Japanese cousin, the N-II. The development costs were Rs4136m (€88m) with the effective cost thereafter of an individual launching €14m (Rs658m). After domestic satellites had been launched, it was ISRO's intention that the PSLV would capture at least some of the lucrative world launcher market.

India' launchers

1980	SLV	Satellite Launch Vehicle.
		17 tonnes, 50 kg into orbit
1987	ASLV	Augmented Satellite Launch Vehicle.
		39 tonnes, 150 kg to LEO
1993	PSLV	Polar Satellite Launch Vehicle.
		275 tonnes, 1000 kg to 900 km polar orbit
2000	GSLV	Geostationary Launch Vehicle.
		402 tonnes, 2.5 tonnes to GEO—due 2000

The size of the PSLV made it the third-largest *solid* fuel rocket in the world (after the boosters on the American space shuttle and the *Titan*). India has claimed to be a world leader in the development of solid fuel rocket engines. India has obtained global patents for its work in the development of caster-oil-based fuels, India being the second largest producer of castor oil in the world.

Rather than construct one giant solid rocket stage, the manufacturers made five smaller 2.8 m diameter segments, each 25 tonnes in weight. The first stage generated 45 tonnes thrust for 90 sec. The fuel used was hydroxyl-terminated polybutadiene resin, with, as oxidizer, ammonium perchlorate. The fuel was fortified by high-explosive to give extra performance. Take-off of the PSLV was assisted by solid-propellant strap-on rockets, two of which ignited on the pad and four after 30 sec. The first stage weighed 128 tonnes and was made of maraging steel. The solid fuels were entirely indigenously developed. The first stage of the PSLV was first tested in October 1989. Placed horizontally near the launch site at Sriharikota, the rocket gushed a blast of flame twice its own length, sending enormous dirty brown clouds billowing into the southeast Indian sky.

The third stage was also solid-fuelled but was made of different material—polyamide kevlar. During ascent, course was maintained by two independent packages of thrusters. The third stage was India's most advanced solid propellant rocket with a specific impulse of 293 sec.

The PSLV was an unusual mixture of solid- and liquid-fuelled rocket. Until that point, rockets had been either liquid- or solid-fuelled, or had been liquid-fuelled with relatively small solid strap-ons. Although Indian rockets had specialized in solid fuels, India had in fact tested its first liquid-fuelled rockets many years earlier—for the first time on 15th May 1974. A liquid-fuelled second stage, measuring 25 cm in diameter and generating 600 kg thrust, had been put on the second stage of a sounding rocket.

The PSLV now marked the introduction of liquid-fuelled rockets in the Indian space programme. The second and the fourth stage had storable liquid propellants—UDMH (unsymmetrical dimethyl methyl hydrazine) and nitrogen tetroxide. The second stage used a European motor, the SEP (the French Societé Européenne de Propulsion) Viking, the same one as used on the European *Ariane* programme and which generated thrust for 145 sec. This required the construction of new test facilities, namely the Liquid Propulsion Centre at Mahendragiri. The first Vikas engines were tested there in 1988.

The fourth stage, built indigenously, had two identical engines which burned for 425 sec, its propellant tanks being built of titanium, with engines that could be gimballed in two planes. Here the fuels were slightly different—mixed oxide of nitrogen (3% MON-3/97%N_2O_4) and monomethyl hydrazine (MMH).

LAUNCHING THE PSLV

The first PSLV launched on 20th September 1993. At first, all went well. The PSLV rose steadily, small items of débris flying free at liftoff, all as scheduled. But staging problems between second-stage cut-off and third-stage ignition led to the third stage being at the wrong angle at ignition. By the time of fourth stage ignition, the top of the rocket had reached an altitude of only 340 km. The fourth stage lacked sufficient thrust to get the payload into orbit and it fell back to Earth. A failure analysis committee was set up and determined that the problem was due to 'a software error in the pitch and control loop of

PSLV launch sequence (sec)	
0	First-stage ignition
0.4	Ignition of two solid strap-ons
29.4	Ignition of four solid strap-ons
74.4	Separation of first two strap-ons
89.4	Separation of four solid strap-ons
105	First-stage separation, second-stage ignition
261	Second-stage separation
262	Third-stage ignition
367	Third-stage separation
640	Fourth-stage ignition
1050	Cut-off
1070	Release of satellite

the on-board guidance and control processor which occurs only when the control command exceeded the specified maximum limiting value' and an unintended contact between the second and third stage. It determined that the rocket's design was fundamentally sound. However, the expensive and valuable IRS payload was lost: the satellite was in fact the refurbished engineering model of IRS-1A. It carried a German Monocular Electro-Optical Stereo Scanner, a LISS-1 scanner and a carbon dioxide monitor.

On the next occasion there were no such problems and on 15th October 1994, the PSLV put into 825-km polar orbit a 870-kg remote sensing satellite, the IRS-P2. The rocket was painted dark red and white, with the Indian flag atop the shroud and the letters 'India' painted on vertically below the ISRO logo. In order to avoid overflying Sri Lanka, achieving polar orbit required controllers to steer the rocket through a 55° yaw manoeuvre 100 sec after liftoff.

INTRODUCING IRS-POLAR

The P series of IRS was successfully introduced with the first polar SLV on 15th October 1994. Centrepiece of the new satellite was the new LISS scanner, LISS-II, designed by the Space Applications Centre in Ahmedabad, Gujarat, with a swath of 131 km. The CCD camera operated in visible light and infrared, with a resolution of 30-m across four spectral bands. The desk-sized spacecraft had two solar panels providing 510 W of electrical power. The mission focused on mapping water resources, checking for floods and monitoring agricultural yields. Under the government's 'Integrated mission for sustainable development', 1992, the country was divided into 157 zones for water management and the satellite helped build up a water picture for the different areas. The concept of 'integrated' was that the programme brought together space and ground-based sensing; diverse disciplines such as public administration, health, science, and soil development; and was linked to programmes of rural development. IRS P2 returned 60 000 images in its first year of operation.

In its second successful launch, the PSLV put another remote-sensing satellite, IRS-P3 into orbit on 21st March 1996. IRS P3 concentrated on the assessment of vegetation,

	PSLV	

Weight: 275 tonnes
Length: 44.2 m
Diameter: 3.2 m
Performance: 1000 kg to 900 km orbit

Stage	Engines	Thrust, kg
Six strap-ons	S9 solid	361 000
1	S139 solid	495 600
2	Vikas liquid, UDMH	60 000
3	S7, solid	33 500
4	Two LUS liquid, MMH	1 400

snow studies, geological mapping for minerals and the analysis of chlorophyll in the oceans. P3 carried an X-ray astronomy telescope, with three counters, a sky monitor, a pinhole camera, and a German modular opto-electrical scanner. P-3 carried a wide field sensor with an image area of over 800 km and a resolution of 190 m, able to provide broad-scale maps of India.

On its fourth flight, the PSLV introduced the C, or continuation, version of the launcher. This had ten tonnes more propellant in the first stage (139 tonnes of hydroxy-terminated polybutadiene) and 2.5 tonnes more propellant in the second, increasing capacity by 280 kg. The second version increased payload from 930 kg to 1250 kg.

IRS-1D IN TROUBLE—BUT SAVED

Watched by the prime minister, the brown-and-white PSLV rose into the September sky of Sriharikota on 28th September 1997. The national flag was mounted on top, flanked by an emblem to mark 50 years of Indian independence. IRS-1D was a replacement for IRS-1C and carried a panchromatic camera, linear imaging self-scanner and wide-field sensor, fed by six solar panels, and had experimental facilities to measure its orbit in reference to the Global Positioning System. It was aimed at a circular orbit at 820 km.

There was a thrust shortfall of 130 fps in the fourth stage due to what appeared to be a fuel tank leak. Controllers in Sriharikota noted that the rocket was not achieving the intended velocity and commanded the engine to burn another 15 sec when the final 420-sec burn ended.

The 300-km perigee was too low for imaging and worse, so low as to lead the IRS to spiral into the Earth's atmosphere and destruction in a few months. In an attempt to save the mission, the satellite used its thrusters to gradually raise its orbit, in the knowledge that the satellite would not be able to carry out its full three years of operations planned. The setback came at a bad time, for ground control in Hassan was already struggling with power failure on the INSAT 2D. The chairman of ISRO, Dr K. Kasturirangan went straight from Sriharikota to the ISRO tracking centre in Bangalore to join his colleagues

The new PSLV on the pad at Sriharikota

in emergency session to save IRS. The satellite had 84 kg of fuel, originally designed for station-keeping over the three years of the satellite's normal operation. By 3rd October, ISTRAC had fired the 11-kN thruster engine five times and had raised the perigee to 375 km and planned to use small burns to raise the orbit to something in the range of 600 km to 680 km. By mid-month, they had raised it to 700 km and the cameras on board the spacecraft were working perfectly. By December, ISRO was able to announce that the satellite would be able to complete a five-year programme of observations.

FIRST OCEAN RESOURCES SATELLITE

On a clear morning, cheered on by Indian Prime Minister Atal Bihari Vajpayee, India's first dedicated ocean observation satellite was put in orbit by the PSLV on 25th May 1999. IRS P-4, more popularly known as *Oceansat 1*, with a weight of 1050 kg, carried an ocean colour monitor and a multi-frequency scanning radiometer able to penetrate clouds. Power came from two 9.6-m^2 solar arrays generating power of 750 W.

Flying piggyback were a 107-kg remote sensing and plasma satellite from South Korea called KITSAT and a 45-kg experimental remote sensing German satellite called TUBSAT. It was India's first launch of foreign satellites, for which they were paid Rs44m (€943,000). By the sixteenth orbit, the main P-4 systems had been deployed and signals were received in Bangalore, Sriharikota, Lucknow, Mauritius, Biak (Indonesia) and Bear's Lake (Russia). By the seventeenth orbit, the camera had been activated and soon provided a rich colour image of southern India, the blue Bay of Bengal and the northwest of Sri Lanka. *Oceansat* was the first Indian marine observation satellite. Its orbit was fine-tuned in early June with the firing of 11-N thrusters to settle the spacecraft in a perfect 727-km orbit crossing the equator at precisely noon local time every day.

P-4's launch had been delayed by American sanctions following India's nuclear tests of 1998 which had prevented the fitting out of the ocean colour monitor, but the equipment had, in the event, been obtained from Germany. The ocean colour monitor operated in eight spectral bands in a 1420-km swath and was used to detect chlorophyll, plankton, aerosols and sediments. The monitor was capable of covering the country every two days.

FUTURE EARTH RESOURCES SATELLITES

IRS P-5 continues the path of specialization developed in the second IRS series. IRS P-5 is known as *Cartosat* and is a mapping satellite with stereo cameras fore and aft. It will

---

IRS in summary

IRS-1A	17 Mar 1988	Baikonour	*Vostok*
IRS-1B	29 Aug 1991	Baikonour	*Vostok*
IRS-P2	15 Oct 1994	Sriharikota	PSLV
IRS-IC	28 Dec 1995	Baikonour	*Molyniya*
IRS-P3	21 Mar 1996	Sriharikota	PSLV
IRS-1D	28 Sep 1997	Sriharikota	PSLV
IRS-P4	25 May 1999	Sriharikota	PSLV

Scheduled

IRS-P5	*Cartosat*
IRS-P6	Agriculture and water resource (Resourcesat)
IRS-P7	Oceans and fishing
IRS-P8	Atmosphere and environment
IRS-3	Introduction of Synthetic Aperture Radar

IRS-1D: rescued

have a wide-field camera, LISS 4 scanner with 6-m resolution and LISS-3 scanner with 23-m resolution, all designed to assist in the detailed mapping of India from 617-km polar orbit. IRS P-6, called *Resourcesat*, will be dedicated to the study of vegetation, P-7 to the oceans and fishing, P-8 to the atmosphere and environment. IRS-P7 will have a scatterometer and an altimeter to measure changes in sea height. All will use the PSLV. Later, the third generation IRS 3 will carry the first Indian Synthetic Aperture Radar.

FIRST INDIAN IN SPACE

The idea of an Indian cosmonaut had been first suggested casually by the world's first space traveller, Yuri Gagarin, when he visited India in 1961. In 1978, the Soviet government made the suggestion of a guest Indian cosmonaut to prime minister Moraji Desai, with a formal agreement subsequently being reached between Leonid Brezhnev and Indira Gandhi in 1980. The offer was to fly an Indian cosmonaut up to the manned Soviet orbiting space station *Salyut* for a week-long visit. From 1976, cosmonauts from guest nations had been training in the Soviet Union, the first of them flying from 1978 onwards. Initially the guests were from the Soviet bloc countries, but the USSR widened the programme, with a French cosmonaut flying in 1982.

One hundred and fifty people applied for the mission and the Indian authorities cut the number to eight. Eventually six were sent to the Soviet Union for evaluation the following year. Two were recommended by the Russians—Wing Commander Ravish Mulhotra and Squadron Leader Rakesh Sharma. Mulhotra was born 25th December 1943 in Lahore and Sharma on 13th January 1949 in Patiala, Punjab. They had 3400 and 1600 flying hours respectively to their credit.

Ravish Mulhotra had made a career in the air force, had flown in both India's wars against Pakistan (1965 and 1971) and was commander of Bangalore Test Pilot School. Sharma was a MiG jet commander in the Indian Air Force and became a test pilot in the Aircraft and Systems Design establishment in Bangalore.

The first trainee cosmonauts were selected in September 1982 and went to Star Town for training. Indira Gandhi met them there during a visit to Moscow, when they were in the process of setting up home on the thirteenth floor of one of the cosmonaut accommodation blocks.

Their first task was to learn Russian and within six months they were speaking the language and taking their lecture notes in Russian. In June 1983 they were introduced to the Soyuz T trainer and taken to splashdown training at Fedosia on the Black Sea. The trainee cosmonauts returned to India in July both for holidays and to visit the Indian research centres where the experiments for their mission were in preparation. The trainees were assigned to their Soviet crew members in October. The prime crewman was Rakesh Sharma, assigned to fly with Yuri Malashev and Nikolai Rukhavishnikov. The backup crew was Ravish Mulhotra, Anatoli Berezovoi and Georgi Grechko. Although Sharma had less flying experience, he was better at Russian—indeed, he was quite a linguist, being fluent in English, Punjab, Hindi and Telegu. In February 1984, with the actual mission fast approaching, the Russians made a final crew change, pulling Rukhavishnikov because of illness and replacing him with Gennadiy Strekhalov.

A feature of the visiting missions was that the guest-country could fly a number of experiments to be operated by their cosmonaut. The Indians chose 43 experiments. These included:

- The use of the MKF-6M, KATE-140 and hand-held cameras to make photographic observations of the Indian continent in the course of 11 passes by the space station over India during the visit
- Attempts to search for gas and oil (*Terra* experiment)
- Earth resources assessment of the Nicobar, Andaman and Laccadive Islands

- Mapping of the Himalayas and Karakorum regions to assess ice melting
- The use of an Indian-made cardiograph called *Vektor*
- Study of blood flow to the cranial regions
- Study of the vestibular and visual systems (*Anketa* and *Optokinisis*)
- Smelting of silver–germanium spheres and their coating in the airlock (*Evaporator-M*)

In addition, the Indian cosmonaut would be examined three times a day by the doctor on board the *Salyut 7* station, Dr Oleg Atkov. Atkov and his colleagues Leonid Kizim and Vladimir Solovyov were then in the midst of a record-breaking 237-day mission aboard the station.

The experiment which attracted most public attention was a yoga exercise in which the cosmonaut would attempt to test the value of yoga in combating the decay of muscles in space. To complete the Indian theme to the mission, the cosmonaut would bring up typical Indian foods for his colleagues and the permanent crew—principally curry, pine-apple, mango juice, crisp bananas and fruit bars. They would surely be a welcome supplement to the notoriously dull Soviet space food.

The prime crew of Yuri Malashev, Gennadiy Strekhalov and Rakesh Sharma arrived at Baikonour cosmodrome for their flight on 23rd March. Their spacecraft, Soyuz T-11, was rolled out to the launch pad on 1st April and erected into position. On 3rd April, seen simultaneously live on Indian and Soviet television, *Soyuz T-11* soared into a clear blue sky to begin chasing *Salyut 7* across 5000 km of space. After several manoeuvres by *Soyuz*, the two spacecraft were only metres apart by lunchtime the following day. Following docking, Malashev, Strekhalov and Sharma entered *Salyut*, Sharma bringing with him an Indian flag, pictures of the country's political leaders and a handful of soil from the Gandhi memorial. There were now six men aboard *Salyut 7*, its largest ever crew. Sharma was the 139th person in space.

The week of joint Soviet–Indian experiments began the following day. There was a 15-min hookup between Sharma and prime minister Indira Gandhi, held both in English and Hindi, in the course of which Sharma described how beautiful India looked from orbit. Later, Sharma held a press conference in which he responded to questions from Indian journalists at the flight control centre in Moscow. With his Soviet colleagues, he carried out medical and Earth resources experiments. The weather was good over India during his mission, making observations possible and on 8th April the cosmonauts radioed a warning about a forest fire which they had spotted in nearby Burma.

After a week, the cosmonauts loaded up the results of the week's joint mission. On 11th April, after bidding farewell to Kizim, Solovyov and Atkov, the visiting cosmonauts boarded the old *Soyuz T-10* spacecraft (leaving the cosmonauts their own ship), undocked, fired retrorockets and plunged Earthward through the atmosphere. The *Soyuz* cabin came to rest in a ploughed field 56 km east of Arkalyk in the standard recovery zone. They were lifted out of the tiny cabin, seated in special chairs and then flown back to Moscow. Sharma's haul included a thousand complete sets of MKF-6M photographs of India and 200 large-format KATE-140 photographs. He had been in space for 7 days 21 hours.

Sharma returned to test flying in Bombay. His Soviet-made spacesuit was put on exhibition. Sharma broke his leg five years after his famous mission when ejecting from a stalling aircraft but he made a full recovery.

Sharma, Malashev, Strekhalov: crew of *Soyuz T-11*

ASTRONAUTS WHO NEVER FLEW

Rakesh Sharma is the only Indian to have made it into space. There were also plans for an Indian to fly on board the American space shuttle. No sooner was the mission to the *Salyut 7* orbital station accomplished than NASA offered India the opportunity to fly an astronaut on board the American space shuttle. The astronaut would be an engineer or scientist who would assist in the launch of the INSAT-1C satellite from the payload bay of the space shuttle, and like Sharma on *Salyut*, would have the opportunity to bring a set of experimental payloads for testing during the mission. A formal agreement was signed in late 1984, with launch of INSAT 1C set for mid-1986.

Two Indian payload specialists were selected for the mission that would launch INSAT 1C. They were assigned to mission 61J, due to fly on the space shuttle *Challenger* on 27th September 1986, though the final selection of who of the two would fly was never made. The two final candidates were Nagapathi C. Bhat and Paramaswaram Radhakrihsnan. Their fellow crew members would have been Donald Williams, commander; Michael Smith, pilot; mission specialists James Bagian, Bonnie Dunbar and Sony Carter; and a reporter on the first journalist-in-space mission. In the event,

Michael Smith died on the *Challenger* that January and Sony Carter was subsequently killed in a plane crash. It was an ill-fated mission.

Nagapathi Bhat was born on 1st January 1948 in Sirsi, North Kanara, and was a graduate from Karnataka University (Bachelor of Mechanical Engineering, 1970) and the Indian Institute in Bangalore (Engineering Design, 1972). He joined the Indian Space Research Organization in Bangalore in 1973 and worked on the *Aryabhata*, *Bhaskhara* and APPLE projects. Paramaswaram Radhakrishnan was born on 2nd October 1943 in Trivandrum, Kerala, and obtained degrees from University College Trivandrum (Bachelor of Science in Physics and Mathematics, 1963, and Master of Science in Physics, 1965). He also joined ISRO, working at the Vikram Sarabhai centre in Trivandrum on the *Aryabhata*, *Rohini*, APPLE and INSAT programmes, becoming a member of the programme planning and evaluation group of the centre. Later, he became group director of electronics there and head of the test and evaluation division of the systems reliability group.

India' astronaut assignments
Soyuz T-11
Rakesh Sharma (flew)
Ravish Mulhotra

Shuttle, 1986
Nagapathi Bhat
P. Radhakrishnan

With the explosion of the shuttle *Challenger*, the INSAT 1C satellite was reallocated to Europe's *Ariane* rocket and sent into orbit on *Ariane* V-24 on 21st July 1988, being subsequently stationed at 93.5°E. However, not long after launch, the satellite lost a quarter of its power owing to the failure of a diode. Although the weather observation platform was kept fully going, only half the telecommunications systems could be used. In November 1989 the satellite lost attitude and had to be abandoned.

The reassignment also ended the chances of Bhat or Radhakrishnan ever making it into space. Similarly to a number of other payload specialists hoping to fly at the time, their missions were effectively disbanded as the shuttle programme was reorganized. Both were training in Houston, Texas, on the very day that *Challenger* blew up.

CONCLUSION: THE PROGRAMME MATURES

In the late 1980s, India developed an uprated version of it small *Scout*-class SLV launcher to place new, heavier payloads in orbit. Although the Augmented Satellite Vehicle (ASLV) did not make a successful mission till 1992, it eventually placed two stretched *Rohini* satellites into orbit. The big leap forward took place with the Polar Satellite Launch Vehicle, which was ten times bigger than the original SLV. The PSLV continued the regrettably well-established Indian tradition of first flight failures, but quickly went on to become a successful launcher. By 2000, the PSLV had enabled India to develop its own capacity to put large polar orbiting resource satellites weighing more than a tonne into orbital pathways sweeping across the Indian subcontinent.

11

Indian space facilities

India has one of the fastest-growing space budgets in the world and its space industry now employs 17 000 people. This chapter reviews the infrastructure which underpins the national space effort—structure, budget, space centres, tracking systems, launch facilities—and looks at some of the benefits accruing to India for its investment. First, its underlying assumptions are examined.

FIRST PRINCIPLES

The Indian space programme has strong economic, social and political overtones. Leaders of the Indian space programme emphasize that they foresee a distinctive development-focused role for their programme—it is not intended as a pale imitation of the programmes of the richer, developed countries. Indeed, they see the programme as a formidable tool in addressing the problems of development and global economic inequality.

At the 1999 United Nations Conference on the Exploration and Peaceful Uses of Outer Space (UNISPACE for short), the head of the Indian delegation and director of ISRO, Dr K. Kasturirangan, told delegates that space technology had an important role in finding sustainable paths to development, promoting social equity and enabling all citizens to reach a minimum quality of life. Professor U.R. Rao, former director of ISRO, subsequently point out that 25% of the world did not have adequate drinking water and that 40% of its people were illiterate. Half the world has never made a phone call. 84% of mobile phones, 91% of fax machines and 97% of internet access is in the developed world. By 2050, 9 billion of the world's 11 billion population will be in developing countries. 'Space technology must contribute to their health and food', argues Professor Rao. 'Where space technology has been put at the service of sustainable economic development, there has been great success. Satellite television education has liberated women and challenged the old divisions of labour' [12]. 95% of deaths from disaster happen in developing countries: satellites offer these countries the tools of disaster management and the possibility of leapfrogging older technologies to bring health and literacy. At the conference, India formally represented the G77—the group of 77 poorest developing nations pressing the richer countries to support satellite-based systems for Earth resources monitoring, disaster warning and telecommunications.

ISRO space centres
- Vikram Sarabhai Space Centre, Trivandrum (VSSC)
- SHAR centre, Sriharikota Island (SHAR)
- ISRO satellite centre, Bangalore (ISAC)
- Space Applications Centre, Ahmedabad (SAC)
- Liquid Propulsion Systems Centre, Trivandrum (LPSC)
- ISRO Telemetry, Tracking & Command Network, Bangalore (ISTRAC)
- INSAT Master Control Facility, Hassan (MCF)
- Development and Education Communication Unit, Ahmedabad (DECU)

Professor U.R. Rao

ISRO headquarters, Bangalore

ORGANIZATION

The Indian space effort is organized by the Department of Space (DOS), which comes under the prime minister's office. Broad policy-making is determined by the Space Commission, which advises both the department and the prime minister. Indian space research and development operates according to a series of five-year plans, the present being the ninth five-year plan (1998–2002). Flanking the Department of Space are the INSAT coordination committee and the planning committee for the National Natural Resources Management System.

Beneath the Department of Space are five bodies:

- National Remote Sensing Agency
- Physical Research Laboratory
- National Mesosphere–Stratosphere–Troposphere Radar Facility
- Antrix Corporation and
- Indian Space Research Organization (ISRO). ISRO has liaison or branch offices in Bombay and New Delhi

The flagship of the Indian space programme and its largest centre is the Vikram Sarabhai Space Centre, named after the founder of the Indian space programme, located 16 km north of Trivandrum. Geophysically, its location is important for being just south of the magnetic equator. This is the main research and development centre of the Indian space programme, where its rockets are designed. It is the largest ISRO centre, employing over 5000 people. The space centre has sections which deal with avionics, solid propulsion, materials, reliability, planning, computers and information systems. The centre includes the Solid Propulsion Group, which has led the development of India's solid-fuel rockets—although the actual rockets are manufactured at Sriharikota, close to the launch site. Also located there is the Inertial Systems Unit for rocket guidance.

Organization of the Indian space programme

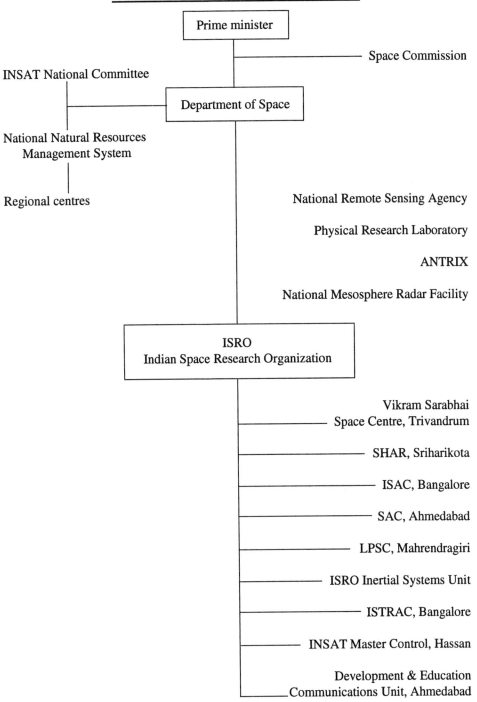

India's space facilities

The sounding rocket launch site at Thumba is part of the centre and was some time ago designated an international and United Nations rocket facility under an agreement with France, Russia and the United States. A wide range of sounding rockets were tested there in the 1960s such as the *Centaure* (France), *Nike Tomahawk* (US), M-100 (Russia) and *Skua* (Britain). At one stage, the M-100 was fired once a week.

The building of the Polar Satellite Launch Vehicle required a major expansion in the facilities in Trivandrum and the construction of an 80-ha extension facility at Valiamala, near Trivandrum. The Rs280m (€5.95m) centre was responsible for computer-aided design, combustion studies, the integration and testing of the PSLV, assisted by the liquid engine testing facility at Mahendragiri and the reinforced plastics centre at Vattiyurkavu.

Many of the most important facilities of the programme are located in Bangalore, one of the main centres for economic and scientific development in modern India, sometimes called India's silicon valley. Bangalore is the headquarters of the Space Commission, ISRO, the Department of Space and the ISRO Telemetry, Tracking and Command. Bangalore is the location of the ISRO Satellite Centre (ISAC) where most Indian satellites have been designed and built. The satellite centre has over 2400 staff. The centre includes the Large Space Simulation Chamber, one of only eight in the world, built in 1990 to simulate the heat, vacuum and solar radiation of Earth orbit. There, satellites are baked, radiated, frozen and vacuumed for period of up to 20 days in a stainless-steel chamber. ISAC also has a blue-walled anechoic chamber.

Further west of Bangalore is Hassan in Karnataka, the Master Control Facility for the INSAT communication satellites. The centre began life with two 14-m dishes for INSAT

1 and then added two 11-m dishes for INSAT 2. Large Earth stations are located in
Bombay, New Delhi, Madras and Calcutta, with smaller stations in Leh, Jaipur, Patna,
Bhubanshwar, Lucknow, Hyderabad, Ernakalam, Kavoratti, Andaman, Nicobar and
Shilong. Remote terminals were set up in the late 1970s in Srinagar, Jodhpur, Bhuj, Goa,
Minicoy, Pondicherry, Aizawl, Imphal, Kohima, Gangtok and Itanagar. By the late 1990s,
this network had expanded to the point that there were over 280 medium and large
ground stations, 5000 small-aperture terminals and 200 stations able to receive satellite
weather data.

Alwaye, between Hassan and Trivandrum, is the site of the Ammonium Perchlorate
Experimental Plant where the fuel for the PSLV is made. It is one of eight solid-fuel
rocket factories in the world and its initial production capacity was 125 tonnes of fuel a
year.

Liquid-propulsion systems are tested in Mahendragiri, near the south coast at a site
nestled under cloud-covered mountains. The centre was constructed for the PSLV and is
called the Liquid Propulsion Test Facility. The services there include: a test stand able to
fire engines for the full duration of their intended burn, including cryogenic engines; a
high-altitude test facility to simulate high altitudes for upper stages; clean rooms;
propellant storage facilities and instrumentation services. Mahendragiri has hydrogen and
monomethyl hydrazine production facilities. The centre employs 1500 people, with offices in Bangalore and Trivandrum. Its main projects are the Vikas stage of the PSLV, the
cryogenic upper stage project and the liquid apogee motor.

Ahmedabad, in the west of India, is the site of the Space Applications Centre (SAC)
the Physical Research Laboratory and the Development and Educational Communication
Unit (the SAC was in practice the bringing together of a range of existing primitive
facilities set up in the late 1960s). The Space Applications Centre is the main applications
centre for the work of ISRO, employing over two thousand people on a lush green office
site. Its first director, appointed in 1973, was Professor Yash Pal. SAC developed the use
of the Indian space programme for telecommunications, television broadcasting, the
survey of natural resources, space meteorology and geodesy as well as the development
of transponders for satellites, portable antennas, dishes, modems and the use of satellites
for news gathering. The Development and Educational Communications Unit is
responsible for the on-the-ground application of television programme for education and
literacy. The Physical Research Laboratory there is responsible for the research
programme in space sciences.

> **Directors of Space Applications Centre**
> 1973–81   Prof. Yash Pal
> 1981–85   Prof. E.V. Chitnis
> 1985–86   Prof. P.D. Bhavsar
> 1986–87   Shri N. Pant
> 1987–94   Shri P.P. Kale
> 1994–   Dr George Joseph

Hyderabad is the headquarters of the National Remote Sensing Agency which operates independently under the Department of Space. The agency has operated the country's main Earth station 55 km away in Shadnagar, which received data from American satellites until India's own *Bhaskhara* and IRS satellites got into orbit. Within the agency there is the Data Centre which archives the information from IRS and distributes products both photographically and digitally. The agency has important roles in interpretation skills and training remote-sensing specialists. The agency operates the *Landsat* ground station and the Indian Institute of Remote Sensing at Dehradun. From 1987, regional applications centres have been set up in Bangalore, Nagpur, Jodhpur, Kharagpur and Dehradun. 23 such regional centres existed by the new century. In 1999, the Indian government approved the setting up of a regional space applications centre for the north-eastern Indian states which border Burma and which can be reached only by a narrow corridor of land to the north of Bangladesh. It will be located in Shilong.

ISRO's tracking system is called ISTRAC. Almost 500 people are employed in the tracking system, the main stations being at Trivandrum, Sriharikota, Car Nicobar, Lucknow and Bangalore, each being equipped with tracking dishes of between 8 m and 10 m diameter. An overseas station is located in Mauritius. The main tracking centre is in Bangalore, called the Spacecraft Control Centre, which has a main mission control room, data and computer facilities. An optical and laser tracking station was built at Kavalur, Tamil Nadu with Russian help. This can pick up satellites as dim as 13th magnitude in Earth orbit as far out as the 24-hr position, 36 000 km high.

Atmospheric science research is carried out at the National Mesosphere–Stratosphere–Troposphere Radar Facility at Gadankinear Tirupati.

Master control, Hassan

SANDBAR LAUNCH SITE: SRIHARIKOTA

The main launching centre is the SHAR Centre on Sriharikota island, 100 km northeast of Madras in Nellore, Andhra Pradesh. Not only is it the principal launching centre, but it is also the main location for static tests. Minor launch sites are Thumba on the southwest coast and Belasore, on the northeast, used for meteorological and sounding rockets. There is a military launching site at Wheeler island on the Orissa coast which in 1999 was used to test the *Agni II* missile during the stand-off with Pakistan.

While close to the sea, like the Japanese launch sites, by contrast the terrain at Sriharikota is flat, covered with eucalyptus, cashew, coconut and casuarina trees in forest, shrub and plantation. From October to November, the site is often lashed by monsoon rains. About 2400 people work at the 145-km² site. The island is shaped like a sand spit, not unlike parts of the Florida coast near the Kennedy Space Centre, another similarity being the narrow channels which run down the middle of the island. The first rockets, sounding probes, were fired from Sriharikota in October 1971.

As one arrives at the launch centre, driving south, there is the security main gate. Soon thereafter is the control centre. The road then branches into two. 7 km to the right is the SLV, later the ASLV complex (now decommissioned). 5.6 km to the left is the PSLV complex and the clean rooms to receive spacecraft before launch. At the landward side of the island are storage centres, administration and accommodation and in the middle is a radio receiving station. A new launch site was built for the GSLV in the mid-1990s.

The island also houses a solid-propellant plant and a static test and evaluation complex. The Solid Propellant Space Booster Plant was built in 1977 for the SLV but was later expanded to manage the much larger rockets of the PSLV. The evaluation centre provides facilities for horizontal firing of rockets, a vibration platform to shake equipment up to $100g$, and a high-altitude test facility simulating conditions of near space for the firing of upper stages.

The PSLV complex was commissioned in 1990. It has an enormous 3000-tonne mobile service tower 76 m tall. It is a mammoth structure, made out of galvanized steel, with a 21 m by 23 m cross section, a 60-tonne crane at the top, able to withstand winds of up to 230 km/hr. It has to be repainted every five years.

The normal launch procedure is to stack the launch vehicles vertically at the pad area in the mobile service tower with a clean room at the top so as to protect the satellite from dust and dirt. This mobile tower brings the rocket down to the pad on twin rail tracks on four bogies before launch, crawling at 7.5 m a minute. The PSLV is then placed on a 175-tonne pedestal ready for launch. The tower is then rolled back 180 m for the launch itself. For the final stages of the countdown, the PSLV is connected to a 50-m tall steel-made umbilical tower which provides cooling and electrical power right up to the last moment. At takeoff, deflector ducts take away the flames of the solid rocket motor exhausts and the rocket rises. The launch control centre is a safe 5.6 km distant.

Before being moved to the clean room on the mobile service tower, satellites are checked out in two clean rooms where they can be examined, balanced, tested for leaks, fuelled and checks can be made of the solar panels and electrical systems.

The first clean room, where the spacecraft arrives, is 7 km from the launch pad, close to mission control. The purpose of this stage is to verify the health of the spacecraft,

PSLV leaves Sriharikota on a pillar of flame

which is kept at a temperature of 22°C with 50% humidity and with less than one dirt particle per million. The second clean room is designed for propellant loading. The third clean room, on the pad, also has an airlock to protect the integrity of the satellite.

Launch tracking stations are located in New Delhi, Kavalur, on the southeast coast, Sriharikota, Ahmedabad, Trivandrum and downrange in the Bay of Bengal at Car Nicobar. As it ascends, the PSLV is tracked by a monopulse C-band precision radar, a dish antenna on a rotating structure which can follow the rocket up to a range of 3200 km, marking its range, height and angle.

Indian sounding rockets

Name	Weight (kg)	Altitude (km)	Payload (kg)	Purpose
Rohini 200	108	80	10	Weather
Rohini 300	370	140	50	Middle atmosphere
Rohini 300 II	504	160	60	Lower ionosphere
Rohini 560	1350	350	100	Ionosphere

SOUNDING ROCKETS FOR SCIENCE, WEATHER

Like many other space programmes, India's started with sounding rockets. They are often forgotten as national space programmes move on to more prestigious activities. Nevertheless, sounding rockets continue to play an important role in weather forecasting and scientific studies and India continues to launch almost one a week. Thumba has three launch pads, capable of taking sounding rockets up to 56 cm in diameter. Sounding rockets continued to be fired from there throughout the 1990s.

India currently has three sounding rockets, which launch not only from the original sounding rocket site of Thumba near Trivandrum but also from Belasore in Orissa and Sriharikota. The rockets are built at the Vikram Sarabhai Space Centre. All are solid-fuelled. The 200 and 560 have two stages whereas the 300 series have only one. They shoot into the atmosphere at enormous speed—indeed, the Rohini 200 has a burn time of less than 8 sec. The Rohini 560 is much the largest and heaviest and is a red-painted rocket fired at an angle from a crane-like platform.

In 1993, a sounding rocket campaign was run in Sriharikota. Two RH-560 were launched only 30 min apart, each releasing barium clouds at altitudes of up to 300 km to measure the distribution of ions and electrons in the ionosphere. A third RH-560 was fired later for the same purpose with German instruments. A Rohini sounding rocket later sent a 127-kg payload up to 432 km on 29th April 1998. A longer, heavier Rohini 560 mark II is in development.

SPREADING THE BENEFITS TO INDUSTRY

With its emphasis on applications, it is no surprise that India chose to emphasize the importance of spin-off and the transfer of the benefits of the space programme to Indian industry. A technology transfer coordinator was appointed at ISRO headquarters in Bangalore, a manage of technology transfer was appointed at the Vikram Sarabhai Space Centre in Trivandrum and other officials with responsibility to promote spin-off were put in place at the National Remote Sensing Agency in Hyderabad, at the Space Applications Centre in Ahmedabad, at Sriharikota and the Liquid Propulsion Systems Centre. By 1998, 231 distinct new technologies had been licensed to private industry.

Much of the spin-off reflects the orientation of the Indian space programme toward Earth resources, remote sensing, weather forecasting and telecommunications. Specific examples of spin-off technologies include mapping systems and cameras, communications terminals, colour printers, film coatings, graphic displays, antennas, water-measuring meters and systems for news-gathering and dissemination.

> **Indian space programme—main areas of technology spin-off to industry**
> Chemicals
> Telecommunications
> Computer systems (hardware and software)
> Electronics
> Optics
> Mechanical systems
> Television
> Ignition systems

In addition, there is a broad area of spin-off deriving from the construction of rockets, satellites and their related ground facilities. Examples are adhesives, glass and carbon composites, lubricants and chemicals deriving from solid rocket fuels. Some of the spin-off items are apparently mundane, but nonetheless important, such as crane controls (deriving from building launch towers), domestic electric shock protectors and even (in the 1987 spin-off list) a 'worm-driven mechanical jack'. A new composite material developed for rockets called kevlar-reinforced polymetamethacrylate was developed as a cheap, tough, non-toxic alternative to traditional dentures.

Since much of the hardware used in the Indian space programme is contracted out to private companies, there are considerable gains to know-how in Indian industry. About 500 companies participate in the space effort. The construction of rockets and satellites involves the manufacturing of products to unprecedently high levels of strength, precision and reliability. Even the making of such mundane items as batteries becomes a challenge when they must work unaided for years, otherwise the whole satellite will fail. The need for high standards is especially true in the making of solid rocket boosters, where India has developed world-class expertise in the making, mixing and use of solid rocket fuels and the forging of steel for casings, joints, rings and segments. This has involved the mastery of a range of materials—maraging steel, aluminium, copper, titanium, cobalt and magnesium. The recent building of a domestic liquid hydrogen engine involves the management not just of highly explosive fuels under great pressures, but control of hydrogen at temperatures of −200°C, a feat achieved elsewhere by few other industrialized nations.

INDIA'S SPACE BUDGET

India's space programme can claim to be one of the most cost-effective in the world. It has been careful to limit and focus its ambitions and not indulge in prestigious projects. It has obtained considerable and often generous help from abroad, either at nil or low cost, as may be seen from such examples as the use of the *Symphonie* satellite (France), ATS-6 (the United States), the APPLE mission (Europe) and several free launches from the Soviet Union. Indian space spending has been small compared to its neighbour China. The space programme has never been politically controversial, having obtained the support of all the prime ministers from Nehru onward, and has rarely been criticized in Parliament.

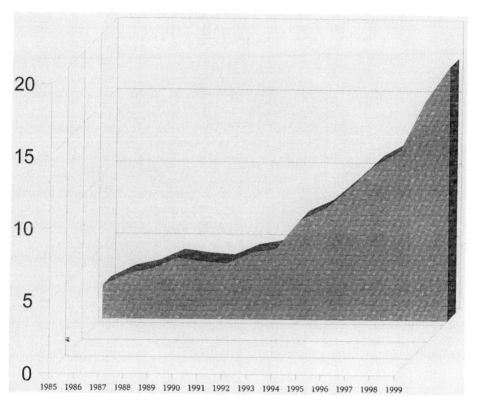

Indian space budget, 1985–89, in billion rupees

The Indian space budget was less than Rs5bn until 1993. Thereafter it began to climb rapidly, passing the Rs10bn mark in 1996 and currently, with a recent steep increase, standing at Rs17.598bn (€374m), though allowance must be made for inflation.

Turning to the details of the budget, the two main headings are satellite operations and launcher development. Launcher development is necessarily an expensive item, with the current development of the new GSLV launcher (Chapter 12). The emphasis on satellite operations is explained by the importance of having a sufficient ground infrastructure to operate satellites and process the incoming results.

The emphasis on space applications is the legacy of the vision of Indian space development set out in the 1970s. Although science is a low priority within the programme, it

Indian Space budget 1999–2000, bn Rs	
Admin.	1.0466
Launcher development	6.4681
Satellite development	0.85554
Applications	1.4295
Science	0.3594
Satellite operations	7.4392
Total	*17.5988*

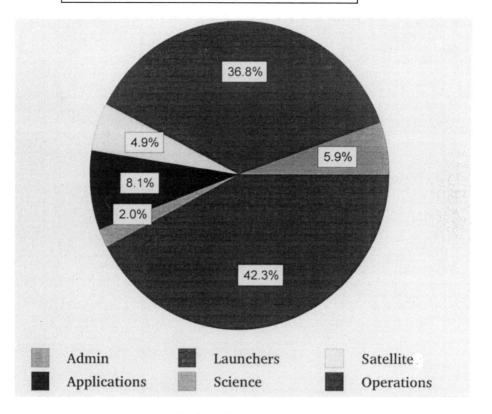

Internal allocation, Indian space budget, 1999

is far from absent, research being carried out at the Physical Research Laboratory in Ahmedabad, the Space Physics Laboratory in Trivandrum and at the National Mesosphere–Stratosphere–Troposphere Radar Facility at Tirupati. Small space science instruments were carried on SROSS-C2 and IRS-P3. There is evidence that, at one stage, India was tempted by some of the more prestigious and adventurous missions of the type mounted by Japan: consideration was given in 1992 to a mission to Mercury using the new GSLV, with the possibility of further, follow-on flights to Venus and Mars. Some diagrams were even published [13] but the proposal was never seriously developed.

CONCLUSIONS: A MODERN INFRASTRUCTURE

India has the full range of infrastructural facilities expected of a modern spacefaring nation, with a national launch centre, tracking systems, space development centre and specialized facilities for sounding rockets, space applications, liquid propulsion and education. The spin-off from space development has been of considerable benefit to industry.

12

Future prospects

The main challenge for the Indian space programme at present is the development of a powerful rocket launcher able to place satellites in geosynchronous equatorial orbit. This will give India the full launcher independence it has always sought.

GSLV CONCEPT: *GRAMSAT* TO THE VILLAGES

The Indian space programme had always concentrated on applications satellites, designed to relay communications, provide weather forecasting and carry out Earth resources surveys for the subcontinent. Although such work can be carried out effectively by small satellites in low Earth orbit, some of these needs are more efficiently met by satellites in 24-hr orbit. India's domestic launchers were able, at best, to put relatively small satellites into low Earth orbits. To reach 24-hr orbit, India had relied on other countries. From 1982 to 1997, India orbited nine INSAT communications satellites. Three flew from Cape Canaveral—two on a *Delta* and one on the space shuttle. Six flew from Guyana on the European *Ariane* launcher. Although they more than repaid their investment, commercial foreign launches were expensive.

Accordingly, India defined in 1987–88 the need for a domestic rocket able to put its own payloads into 24-hr orbit. It was named the Geostationary Satellite Launch Vehicle (GSLV). The original aim of the GSLV was that it would carry the INSAT 2 series and that it would place these satellites into orbit at much lower overall cost than foreign launchers—the aim was to reduce launch costs from €56.6m (Rs2.66bn) on *Delta* and *Ariane* to an effective cost of €18.8m (Rs883m). However, the development costs of the GSLV were substantial, being estimated at Rs13bn (€276m). One factor in India's favour was that in Sriharikota it had a launching range relatively near to the equator.

A specific payload intended for the GSLV was *Gramsat*, an educational relay platform based on the INSAT 2 design (*gram* is the Hindi word for village). Based on the INSAT 2, *Gramsat* was designed to transmit television, CD-quality sound, data and internet to 60-cm dishes. *Gramsat* would have a national beam with six powerful transponders for educational television for rural development, transmitting programmes in health and hygiene, agricultural production, family planning, environmental awareness, vocational training and entertainment. It would also have two advanced spot beams for village education programmes, using one visual signal but able to transmit sound in four different languages. Ion engines would be used for station-keeping.

For the bottom stages of the GSLV, India decided to take advantage of its existing knowledge of the behaviour of rockets in the lower atmosphere and rely on the existing polar SLV. The new rocket would use the PSLV's solid-fuelled first stage, but instead of using solid-fuel strap-ons; it would use four PSLV second-stage liquid-fuelled engines as strap-ons, thereby generating enormous thrust at liftoff. The PSLV Vikas second stage would be used as the GSLV second stage as well. First ground tests of the GSLV's engines were made at Mahendragiri in June 1994. For the very final, fourth stage of placing payloads into geosynchronous orbit, the GSLV would use the Indian Liquid Apogee Motor or LAN, a 440 N liquid propulsion system which had already accumulated 10 000 sec of testing.

Where the launcher broke new ground was the concept of a cryogenic, liquid-hydrogen-powered third stage. Cryogenic fuels, while enormously powerful, require considerable handling skills since the boiling point for liquid hydrogen is −253°C and −183°C for liquid oxygen. The concept of an Indian cryogenic upper stage was approved by DOS in late 1987. An engine of about 12 tonnes would be required.

At that stage, India was in the process of developing a one-tonne cryogenic engine but was not confident about bringing the project to fruition. Indian engineers calculated that up to 15 years' work was involved to fully master the difficult technology. Originally, India made inquiries with Japan about purchasing the LE-5, but nothing seems to have come of this [14]. The Indians were then approached by the General Dynamics Corporation in the United States offering both a cryogenic engine and technology transfer. The cost was prohibitive and ISRO decided, with the approval of the Space Commission, to reconsider its decision to shop abroad and to go it alone, despite the long time this would take. Also aware of India's interest in cryogenic technology, Europe's Arianespace made an approach, but again at a prohibitive cost. This was becoming a highly political matter and ISRO went to the cabinet to confirm a go-it-alone decision. Just before it did so, a third approach came, this time from the Soviet Union, offering two engines and technology transfer for the more reasonable price of Rs2.35bn (then about €188m).

In 1988, outline agreement was reached between the Indian Space Research Organization (ISRO) for the development of a cryogenic upper stage with Glavcosmos, the marketing agency for the Soviet space programme. The outline agreement was reinforced with a more definitive, technical agreement in June 1991, which also engaged the Soviet authorities for Russia to launch the first *Gramsat*. The main provision was that USSR would deliver two engines and the associated technology by 1995 for the stipulated amount of Rs2.35bn.

KVD-1: REMNANT FROM THE MOON RACE

The Soviet Union offered India the use of a rocket engine not then known in the west, the KVD-1, built by the Isayev design bureau. The KVD-1 engine has a burn time of 800 sec and a combustion chamber pressure of 54.6 atmospheres. The KVD-1 had a turbopump-operated engine with a single fixed-thrust chamber, two gimballed thrust engines, an operating period of up to 7.5 hours and a five times restart capability. It weighed

3.4 tonnes empty and 19 tonnes fuelled. Its thrust was 7300 kg and the specific impulse 461 sec, still the highest in the world by the end of century [15]. It was 2.146 m tall and 128 m in diameter and it weighed 292 kg. This small object was to spark off international controversy.

The Isayev bureau was one of the least well-known of all the Soviet design bureaux and featured little in the early *glasnost* revelations about the Soviet space programme, its design bureaux and rocket engines. The bureau started life as plant 293 in Podlipki in 1943, directed by one of the early Soviet rocket engineers, Alexei M. Isayev (1908–71) and was renamed OKB-2 in 1952, being given its current name, KM KhimMach in 1974.

Besides spacecraft, its work has concentrated on long-range naval, cruise and surface ballistic missiles and nuclear rockets and by the early 1990s had built over 100 rocket engines, mainly small ones for upper stages, mid-course corrections and attitude control.

The KVD-1 was not a new motor—it was originally developed as part of the Soviet-manned Moon landing programme as far back as 1964. The Isayev bureau was then tasked with responsibility for developing a powerful, hydrogen-fuelled upper stage which would support the establishment of long-stay manned lunar missions after the first Soviet manned landings.

What the fuss was all about: the KVD-1 (this shows a later version, the KVD-1M)

Although the USSR was beaten to the Moon, the Isayev rocket was called upon in support of an ambitious plan of lunar exploration to take place following the end of the Apollo programme when the Russians would have a clear run. The N-1/L-3M or GB plan, approved in early 1972, had the objective of putting three cosmonauts on the Moon initially for 5 to 14 days but later for a full lunar month. This mission required the KVD-1, then called the block R and with the industry code of 11D56. Its role was to brake the assembly into lunar orbit and make the descent to the lunar surface. The KVD-1 was first test-fired in June 1967. The engine was tested for 24 000 sec in six starts.

However, the N-1/L-3M programme was cancelled in 1976, so the KVD-1 never flew and went into storage. For subsequent missions to geosynchronous orbit and deep space, the Russians contented themselves with using kerosene/liquid-oxygen-based engine, the 'block D'. The West wrongly presumed that the Russians had not been able to develop a cryogenic upper stage. In fact, the KVD-1 had unsurpassed thrust and capabilities that made it unmatched for years. Five block R stages were built and tested over the years 1974–77—indeed a series of tests were run over the 1967-76 period and the engine was declared fully operational.

AMERICAN REACTION

Development of the Indian upper stage with Russian help had been underway for four years when the arrangements were denounced by American President George Bush as a violation of the Missile Technology Control Régime. In May 1992, the Bush administration announced that it was applying American sanctions on both ISRO and Glavcosmos for two years. The sanctions involved American non-cooperation with both agencies and the introduction of sales embargoes on India. India objected strongly to the American actions, pointing out that high-powered hydrogen-fuelled upper stages which took a long time to prepare were of little military value in attacking a neighbouring country with which they already had a land border. To prove the connexions between space and the military, the Americans even drew up a list of seven ISRO scientists who had gone to work on missiles—as if there was not a similar traffic in the United States. India also pointed out that the Americans had offered them the very same technology themselves and that the Americans never made any objection throughout the years 1988–92 when the arrangements had begun.

In 1993, with the accession of Bill Clinton as president, the American attitude relented. He approved a reopening of cooperation with ISRO and Glavcosmos if the Russians transferred individual engines, but not the production technology that would enable India to design its own cryogenic engines. In July 1993, after negotiations in Washington DC, Russia backed off its contract to transfer technology to India and suspended its agreement (August), invoking *force majeure* (circumstances beyond control), to the Indians' fury.

The KVD-1 had now become caught up in a much bigger game—the negotiations between America and Russia for the construction of the International Space Station. Russia suggested compensation for loss of the Indian contract: the $400m paid by the United States for seven American flights to *Mir* at this time may have become part of the equation.

In a revised agreement with India made in January 1994, Russia agreed to transfer three, later renegotiated by the Indians to seven, KVD-1s intact, without the associated technology. The negotiations were later described as very tough, but the Indians managed to negotiate a much lower price—a figure of €8.5m was indicated—and extra engines in exchange for the loss of technology transfer. In addition, two boilerplate models would be supplied to test how they would best fit the launcher shroud. India was required by the United States to agree to use the equipment purely for peaceful purposes, not to re-export it, nor to modernize it without Russia's consent.

The KVD-1 was not identified in public until 1995, when a model went on show at the Paris exhibition. There, it was presented not only as the upper stage for the Indian GSLV but was given a new lease of life as the upper stage of an uprated *Proton M* rocket and the forthcoming new *Angara* rocket due to enter service early in the twenty-first century (the *Angara* version is called the KVD-1M).

ENDGAME

According to some American sources, the Russians transferred the production technology in any case [16]. The appropriate documents, instruments and equipment were allegedly transferred in four shipments from Moscow to Delhi on covert flights by Ural Airlines. As a cover, they used 'legitimate' transhipments of Indian aircraft technology travelling the other way to Moscow for testing in Russian wind-tunnels. The plot thickened when at the same time in October 1994, two senior scientists at the Liquid Propulsion Systems Centre, S. Nambi Narayanan and P. Sasikumaran were arrested for 'spying for foreign countries'. Eventually, the Central Bureau of Investigation admitted that the charges against them were false and baseless and they were freed. Later, the United States was accused of setting them up as part of a dirty tricks campaign against the sale of the KVD-1.

Finally, two boilerplate models were delivered between 1997 and 2000. The first of seven ready-to-fly KVD-1s arrived on 23rd September 1998 in Madras, whence it was taken to Sriharikota.

There were a number of outcomes to this episode. In the first instance, India's attempts to develop a performance to 24-hr orbit was delayed by at least five years. The restrictions on technology transfer obliged the Indians to develop, certainly sooner than they might have done, their indigenous cryogenic upper stage. The Russians lost at least four-fifths of the price of their original deal. While the Americans may have in part been motivated by concerns of proliferation, it seems that their desire to stop new launcher competitors from entering the global launcher market may have played a more important role.

In the event, the Indians eventually made good progress with their own upper stage, called the CUS, or Cryogenic Upper Stage. India began to develop its own cryogenic technology in the meantime and tested its own sub-scale engine for the first time on 21st July 1989. A hydrogen manufacturing plant was opened in 1992, and a thrust chamber was made that year. Ironically, India obtained its liquid hydrogen supplies from an American company which built a plant for ISRO at its test facility in Mahendragiri. The designers required 1200-sec continuous running before they were prepared to approve the design for flight. However, things went far from smoothly for there was an explosion when India fired a 10-kN liquid hydrogen engine in July 1993. Two others exploded and another gave up after 10 sec. A prototype engine with one tonne of thrust was eventually tested at the Liquid Propulsion Systems Centre in Mahendragiri in 1997.

GSLV		
Length: 51 m		
Diameter: 2.8 m		
Weight: 402 tonnes		
Performance: 2.5 tonnes to geostationary orbit		
Stage	Type	Thrust, kg
Strap-ons	4 liquid-fuelled Vikas	$4 \times 74\,000$
1	Solid	360 000
2	Liquid Vikas	74 000
3	1 KVD-1, later Indian CUS	7 500

The Indian upper stage will have 7500 kg thrust, 461 isp and 740-sec duration. In the course of time, India plans to develop 12 000-kg and 16 000-kg thrust versions. The liquid engine strap-ons of the GSLV, the L-40, were first ground-tested in July 1995. The 60-tonne engine was tested for 200 sec. Meanwhile, following further ground tests in May 1997 and March 1998, the system was declared to be qualified for flight. From 1997, GSLV stages were regular appearances on test stands as the day for the first launch drew closer. Even as it did so, Indian officials had already begun to turn their minds to means of upgrading the GSLV, doubling its payload to 24-hr orbit to five tonnes. Construction of a successor was set for 2005.

COMMERCIALIZATION

The flying of KITSAT and TUBSAT in May 1999 on the PSLV marked India's entry into the international commercial launcher market. Earlier hopes to secure commercial contracts had been dashed when in 1997 the award for the *Iridium* global mobile communications network went to China's *Long March 2*. Under an agreement reached in 1998 between the commercial wing of the Department of Space, Antrix Corporation and Europe's Arianespace, the PSLV was marketed jointly with the European *Ariane* for small payloads up to 100 kg. Soon after, it was agreed that PSLV would fly a small 100-kg Belgian satellite called *Proba* with the next IRS (*Cartosat*) in 2000.

Proba was a small Belgian (Flemish) technology demonstration satellite. Standing for Project for On-Board Astronomy, it is a six-sided cube of 80 cm by 60 cm with five sides covered in solar cells and the sixth holding three scientific experiments. These are a spectrometer of 25m resolution, a radiation recorder and a dust detector. The Indians have charged €763 000 (Rs35m) for the launch.

PSLV second-stage engine

Ultimately, the Indians hope that between them, the PSLV and the GSLV will be attractive commercial prospects in the global launcher market. However, this market has proved notoriously difficult to penetrate. It is a market dominated by the United States and Europe, with small shares for Russia's *Proton* and, until the later 1990s, the occasional Chinese launch. The United States used its power to regulate the world launcher trade, imposing quotas on Russian rocket launchings and doubtful security concerns to freeze out the Chinese altogether. The manner in which the United States voided the Glavcosmos–ISRO deal suggests that the US will be equally vigorous in preventing Indian competition in the future.

CONCLUSIONS: THE CHALLENGE

For India, the next challenge in space is the GSLV. This will mark India's entry to the cryogenic club—an exclusive inner circle of rocket nations, comprising the United States, Europe, Russia, China and Japan. Development of the GSLV has proved to be a tortuous, frustrating process, global political issues delaying development by about five years. Once the GSLV flies, India will have the capacity of launching its own dedicated 24-hr applications satellites as it wishes and, once the development costs are paid off, at much more economical costs than on the global launcher market. The *Gramsat* project opens up exciting new prospects in the use of satellites for education and development. Whether India will make an inroad into the world commercial launcher market is doubtful.

13

Two roads into space

Finally, 'Two roads into space' examines the Japanese and Indian space programmes in a global context, commenting on their distinctive features.

IN GLOBAL CONTEXT

Before placing Japan and India in a global context, it may be useful to compare their two programmes with one another.

	Japan	India
Successful launches	54	9
Failures	8	4
Launches by other countries	19	16
	81	29

Japan's programme is clearly much the larger of the two. Although India has had more satellites launched by other countries, Japan leads overwhelmingly in domestic launches. Japan has had very few launch failures, indeed only one from 1970 to 1998, although not all satellites reached their intended orbit.

The Japanese and Indian space programmes are small in world terms. The following table compares world, Japanese and Indian launchings for the last 15 years of the 20th century.

World launching table: world, Japanese and Indian launch rates, 1957–85 and 1986–99

	1957–85	86	87	88	89	90	91	92	93	94	95	96	97	98	99
World	3811	103	110	116	101	116	88	94	79	87	74	73	86	77	69
Japan	29	2	3	2	2	3	2	1	2	2	2	1	2		
India	3							1		2		1	1	2	1

Japan and India therefore account for only about 2% of global annual launches. It may, however, be more realistic to set India and Japan separately from the superpowers.

Total space launches, 1957–99	
USSR/Russia	2596
US	1188
Europe	117
Japan	54
China	59
India	10
Israel	3

In addition, France carried out 10 launches, Great Britain 1. The French programme of launches became integrated with the European, while the British programme was cancelled. Nine international payloads have been launched from a platform off Kenya and two from a platform off Easter Island.

Japan therefore emerges as the fourth spacefaring nation, or the second in the non-superpower club and may be said to be in the same broad bracket as China. India is very much at the lower end of the Table. Israeli launchings are likely to be occasional. The only new entrant in the immediate future is likely to be Brazil.

SPACE SPENDING COMPARED

India and Japan are therefore small players in the world space business. A more useful indicator may be the amount of money spend on spaceflight.

Compiling such figures is a hazardous exercise, for allowance must be made for a wide range of disparate factors. Japan emerges, in financial terms, as the third individual spacefaring bloc in the world, following the United States and Europe. India is toward the bottom end of the table, with a low-cost but, as we have seen, value-for-money programme. These figures are of course problematical. They do not include China, for which a satifactory contemporary comparable estimate is not available. Russia has a very low current space spending, and is currently spending almost nothing on research and development. On the other hand, it has a huge infrastructure, with development costs paid off years ago and a level of activity far above that of Europe or Japan, so it may be more realistic to place Japan in fourth position. The American figure only includes the civilian space agency, NASA, and not the American military space budget, which is worth the same amount again: then one realizes just how much the United States dwarfs the space budgets of the rest of the world. Some prefer spending per head as a useful indicator and the Japanese Space Agency, NASDA one calculated comparative figures (these have the disadvantage of not including some of the nations included above, but of including Great Britain).

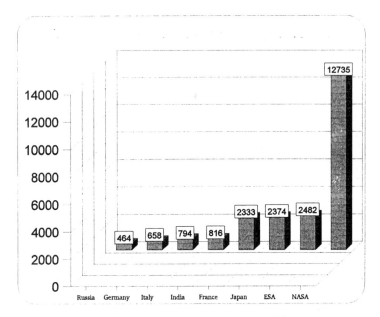

World space budgets, 2001, projected, in million euros. (Source: Adapted from *Air & Cosmos*)

World space spend per taxpayer, ¥/person/year	
USA	¥4,954
France	¥4,060
Japan	¥1,935
Germany	¥1,198
Britain	¥525
Source: NASDA	

These figures show Japan well below United States and French spending, but ahead of Germany and far ahead of Great Britain.

FIELDS OF WORK—JAPAN AND INDIA COMPARED

Turning to the orientation of the two programmes, the programmes may be divided into their respective spheres of work.

Here it may be seen that the balance of activities between the two programmes is quite different. India has concentrated entirely on Earth observations, communications and meteorological satellites. It has launched no dedicated scientific satellites, though some satellites have carried scientific instruments as secondary payloads. By contrast, Japan's programme covers a much broader range of endeavours, with a significant investment in engineering test satellites and scientific missions. While Japan's Earth observations missions have been limited in number, they have used large platforms able to return

Category	Japan		India	
	Domestic	Foreign	Domestic	Foreign
Marine and Earth observations	4		9	6
Engineering	14			
Communications	8	12		10
Meteorology	4	1		
			(See marine and Earth observations)	
Materials processing	1			
Science	19			
Deep space	4			
	54	18		
	72		25	

significant quantities of data of high quality. Japan's deep space programme is especially notable, making it the only country to have such a programme in the late 1990s apart from the United States.

Despite these differences, both programmes have much in common. Both started from small beginnings, using primitive sounding rockets and led by charismatic father-figures, Hideo Itokawa and Vikram Sarabhai. Both began with the placing of small satellites in Earth orbit, using solid-fuelled rockets. Both pursued the path of indigenization—learning how to borrow technology from abroad and rebuild it at home—a path pursued by the Indians with more conviction. The Indian programme has been tightly focused on applications (communications, Earth observations and meteorology) and may be one of thr most concentrated in the world. India has remained faithful to the vision and route mapped out by the founder of its space programme, Vikram Sarabhai. The success of India in indigenization and quality observation work seems to have been vindicated, for India is now the world leader in the use of satellite technology to combat underdevelopment.

With its greater economic resources, Japan has had greater possibilities of developing a more broadly based programme, one incorporating communications, applications, engineering and science. That is not to say that the Japanese programme has lacked focus. To the contrary, Japan has concentrated on programmes of considerable and immediate help to the Japanese islands in the areas of communications, weather warning, marine and Earth observations. Especially impressive has been the investment in engineering test satellites, particularly to develop more sophisticated means of communications. The continuous investment in testing cutting edge technologies shows how Japan has used its programme to successfully develop and maintain its lead in the worldwide development of advanced communications and industrial engineering. No less impressive is the imaginative Japanese scientific programme, not least because of its low cost, the use of micro-technologies and, in the case of deep space missions, its employment of small rockets using idiosyncratic trajectories to fly to the Moon and Mars, demonstrating how much can be achieved with little.

CONCLUSIONS

At the end of the 1990s, the main headlines for the Japanese programme concerned its problems of launcher development and cost containment. While undoubtedly serious, a more positive approach is likely to develop as Japanese participation in the International Space Station gets into its stride and as the *Nozomi*, Lunar A and *Selene* projects come to fruition. India, with its rising space budget and growing launcher capability, will make its mark on the world space industry. For other countries considering their participation in space activities, especially in placing space programmes in the service of development, India sets a high standard to follow.

Annexes

ANNEXE 1: JAPANESE LAUNCHES

Date	Name	Vehicle	Site	Orbit	Inclination	Period
11 Feb 1970	*Ohsumi*	*L-4S*	Kagoshima	338–5,150 km	31.07°	144.2 min
16 Feb 1971	*Tansei*	*Mu-4S*	Kagoshima	990–1,110	29.66°	105.95
28 Sept 1971	*Shinsei*	*Mu-4S*	Kagoshima	869–1,865	32.06°	112.92
19 Aug 1972	*Denpa*	*Mu 4S*	Kagoshima	245–6,291	31.03°	156.85
16 Feb 1974	*Tansei 2*	*Mu 3C*	Kagoshima	284–3,233	31.23°	121.6
24 Feb 1975	*Taiyo*	*Mu 3C*	Kagoshima	249–3,129	31.54°	120.06
9 Sep 1975	ETS-1/*Kiku-1*	N-I	Tanegashima	975–1,103	47°	105.88
29 Feb 1976	*Ume 1*	N-I	Tanegashima	994–1,013	69.67°	105.2
19 Feb 1977	*Tansei 3*	*Mu 3H*	Kagoshima	796–3,821	65.77°	134
23 Feb 1977	ETS-2/*Kiku 2*	N-I	Tanegashima	42,163–35,780	0.1°	1.436
4 Feb 1978	Exos-A/*Kyokko*	*Mu-3H*	Kagoshima	642–3,975	65.09°	134.27
16 Feb 1978	*Ume 2*	N-I	Tanegashima	975–1,224	69.37°	107.25
16 Sep 1978	Exos-B/*Jikiken*	*Mu-3H*	Kagoshima	228–30,051	31.1°	527.17
6 Feb 1979	ECS-1/*Ayame-1*	N-I	Tanegashima	190–34,583	24.09°	607.16
21 Feb 1979	CORSA/*Hakucho*	*Mu-3C*	Kagoshima	541–573	29.9°	95.61
17 Feb 1980	*Tansei 4*	*Mu-3S*	Kagoshima	517–603	38.7°	95.74
22 Feb 1980	ECS-2/*Ayame 2*	N-I	Tanegashima	213–35,592	24.53°	627.04
11 Feb 1981	ETS-4/*Kiku 3*	N-II	Tanegashima	258–36,253	28.61°	641.28
21 Feb 1981	*Hinotori*	*Mu-3S*	Kagoshima	573–639	31.34°	96.65
11 Aug 1981	*Himawari 2*	N-II	Tanegashima	35,776–35,792	0.2°	1435
3 Sep 1982	ETS-3/*Kiku-4*	N-I	Tanegashima	965–1,228	44.62°	107.1
4 Feb 1983	*Sakura 2A*	N-II	Tanegashima	35,145–36,767	0.29°	1445
20 Feb 1983	Astro B/*Tenma*	*Mu-3S*	Kagoshima	488–503	31.49°	94.38
5 Aug 1983	*Sakura 2B*	N-II	Tanegashima	35,535–36,610	0.29°	1451
23 Jan 1984	BS-2A/*Yuri 2A*	N-II	Tanegashima	35,359–36,020	0.4°	1,431
14 Feb 1984	Exo C/*Ohzora*	*Mu-3S*	Kagoshima	356–887	74.59°	97.18
3 Aug 1984	*Himawari 3*	N-II	Tanegashima	35,783–36,340	1.86°	1450
8 Jan 1985	MS-T5/*Sakigake*	*Mu-3S*	Kagoshima	Solar orbit	—	—
19 Aug 1985	Planet-A/*Suisei*	*Mu-3S*	Kagoshima	Solar orbit	—	—
12 Feb 1986	BS-2B/*Yuri 2B*	N-II	Tanegashima	34,406–36,164	0.57°	1410
13 Aug 1986	*Ajisei*	H-I	Tanegashima	1,479–1,479	50.01°	115.66
	Fuji			1,479–1,479	50.012°	115.66
5 Feb 1987	*Ginga*	*Mu-3S*	Kagoshima	510–673	31.08°	96.35
19 Feb 1987	MOS-1A/*Momo*-1A	N-II	Tanegashima	903–917	99.1°	103.27
27 Aug 1987	ETS-5/*Kiku-5*	H-I	Tanegashima	33,535–36,098	0.19°	1,386

19 Feb 1988	*Sakura 3A*	H-I	Tanegashima	34,252–36,565 km	0.07°	1,417 min
16 Sep 1988	*Sakura 3B*	H-I	Tanegashima	35,613–36,583	0.23°	1,451
21 Feb 1989	Exos-D/*Akebono*	*Mu3-SII*	Kagoshima	277–10,460	75.11°	211.23
5 Sep 1989	*Himawari 4*	H-I	Tanegashima	35,720–35,856	1.6°	1,436
24 Jan 1990	Muses-A	*Hiten:*	LOI 15 Feb 1992, impact: 11 Apr 1993			
		Hagoromo	LOI 18 Mar 1990			
7 Feb 1990	MOS-1B/*Momo 1B*	H-I	Tanegashima	911–920	99.08°	103.37
	Orizuru			912–1,746	99.06°	112.28
	Fuji 2			912–1,746	99.06°	112.28
				35,260–37,672	0.3°	1,471
25 Aug 1990	BS-3A/*Yuri 3A*	H-I	Tanegashima	197–37,269	28.55°	659.54
25 Aug 1991	BS-3B/*Yuri 3B*	H-I	Tanegashima	519–787	31.34°	97.62
30 Aug 1991	Solar A/*Yohkoh*	*Mu-3SII*	Kagoshima	567–574	97.69°	96.14
11 Feb 1992	JERS/*Fuyo*	H-I	Tanegashima	538–647	31.3°	96.53
20 Feb 1993	Astro-D/*Asuka*	*Mu-3SII*	Kagoshima	449–459	30.512°	93.66
4 Feb 1994	OREX/*Ryusei*	H-II	Tanegashima	467–36,082	28.54°	642.03
	EP/*Myojo*					
28 Aug 1994	ETS-6/*Kiku 6*	H-II	Tanegashima	7,796–38,707	13.08°	845.96
15 Feb 1995	*Express*	*Mu-3SII*	Kagoshima	110–250	31°	88.09
18 Mar 1995	SFU	H-II	Tanegashima	467–496	28.46°	94.24
	Himawari 5			35,731–35,789	1.06°	1,434
12 Feb 1996	HYFLEX	J-1	Tanegashima	Sub-orbit		
17 Aug 1996	ADEOS-1/*Midori*	H-II	Tanegashima	797–799	98.62°	100.83
	Fuji 3			801–1,323	98.58°	106.46
12 Feb 1997	Muses-B/*Haruka*	*Mu-5*	Kagoshima	573–21,402	31. 32°	379.59
28 Nov 1997	ETS-7/*Kiku-7*	H-I	Tanegashima	379–538	34.97°	93.761
	TRMM			367–385	35°	92.07
21 Feb 1998	COMETS/	H-II	Tanegashima	247–1,883	30.05°	106.51
	Kakehashi					
4 Jul 1998	*Nozomi*	*Mu-5*	Kagoshima	703–489,382	25.4°	22,646

Failures

26 Sep 1966	*Lambda*		Kagoshima
20 Dec 1966	*Lambda*		Kagoshima
13 Apr 1967	*Lambda*		Kagoshima
22 Sep 1969	*Lambda*		Kagoshima
25 Sep 1970	*Mu-4S*		Kagoshima
15 Nov 1999	H-II	MT-SAT	Tanegashima
10 Feb 2000	*Mu-5*	Astro-E	Kagoshima

ANNEXE 2: JAPAN'S LAUNCHES BY OTHER COUNTRIES

14 Jul 1977	GMS-1/*Himawari 1*	*Delta*	Cape Canaveral
15 Dec 1977	*Sakura 1*	*Delta*	Cape Canaveral
7 Apr 1978	BSE-1/*Yuri 1*	*Delta*	Cape Canaveral
6 Mar 1989	JCSat–1	*Ariane*	Korou
5 Jun 1989	*Superbird A*	*Ariane*	Korou
1 Jan 1990	JCSat 2	*Titan 3*	Cape Canaveral
26 Feb 1992	*Superbird B*	*Ariane*	Korou
24 Jul 1992	*Geotail*	*Delta*	Cape Canaveral
1 Dec 1992	*Superbird A-1*	*Ariane*	Korou
9 Jul 1994	BS-3N	*Ariane*	Korou

29 Aug 1995	JCSat-3	*Atlas 2AS*	Cape Canaveral
29 Aug 1995	N-star A	*Ariane*	Korou
5 Feb 1996	N star B	*Ariane*	Korou
17 Apr 1997	BS-4	*Ariane*	Korou
3 Dec 1997	JCSat-4	*Ariane*	Korou
27 Jul 1997	*Superbird C*	*Atlas 2AS*	Cape Canaveral
29 Apr 1998	B-Sat-1B	*Ariane*	Korou
16 Feb 1999	JCSat-6	*Atlas 2AS*	Cape Canaveral

ANNEXE 3: INDIA'S LAUNCHES

18 Jul 1980	*Rohini 1B*	SLV-3	306–919 km	44.75°	96.85 min
31 May 1981	*Rohini 2*	SLV-3	187–418	46.27°	90.49
17 Apr 1983	*Rohini 3*	SLV-3	388–852	46.6°	97.02
20 May 1992	SROSS-C	ASLV	250–424	46.04°	91.17
4 May 1994	SROSS C-2	ASLV	433–922	46.03°	98.3
15 Oct 1994	IRS-P2	PSLV	798–883	98.69°	101.74
21 Mar 1996	IRS-P3	PSLV	818–848	98.8°	101.57
28 Sep 1997	IRS-1D	PSLV	308–822	98.64°	95.96
25 May 1999	IRS-P4/*Oceansat*	PSLV	719–739	98.39°	99.37

Failures

10 Aug 1979	SLV	15 min
24 Mar 1987	ASLV	48 sec
13 Jul 1988	ASLV	48 sec
20 Sep 1993	PSLV	20 min

All launches from Sriharikota.

ANNEXE 4: INDIA'S LAUNCHES BY OTHER COUNTRIES

19 Apr 1975	*Aryabhata*	Kapustin Yar	*Cosmos 3M*
7 Jun 1979	*Bhaskhara 1*	Kapustin Yar	*Cosmos 3M*
19 Jun 1981	APPLE	Korou	*Ariane*
20 Nov 1981	*Bhaskhara 2*	Kapustin Yar	*Cosmos 3M*
10 Apr 1982	INSAT 1A	KSC	*Delta*
30 Aug 1983	INSAT 1B	KSC	Shuttle STS-8
17 Mar 1988	IRS 1A	Baikonour	*Vostok*
21 Jul 1988	INSAT 1C	Korou	*Ariane*
12 Jun 1990	INSAT 1D	KSC	*Delta*
29 Aug 1991	IRS 1B	Baikonour	*Vostok*
10 Jul 1992	INSAT 2A	Korou	*Ariane*
23 Jul 1993	INSAT 2B	Korou	*Ariane*
7 Dec 1995	INSAT 2C	Korou	*Ariane*
28 Dec 1995	IRS C	Baikonour	*Molniya*
4 Jun 1997	INSAT 2D	Korou	*Ariane*
2 Apr 1999	INSAT 2E	Korou	*Ariane*

Bibliography

Numerous information sources are available on the Japanese and Indian space programmes. The American periodical, *Aviation Week* and *Space Technology*, gives full attention to space developments in Japan and India, as does the British *Flight International* and the French *Air & Cosmos*. News and features are published in the British Interplanetary Society's monthly *Spaceflight*.

The Japanese space agency, NASDA, issues a monthly bulletin *NASDA report*. It is available, with a huge range of other material, at its excellent website:

> http://www.nasda.go.jp

ISAS publishes an attractive annual report. Available from:

> 3-1-1 Yoshinodai, Sagamihara, Kanagawa 229-8510

Its website is:

> http://www.isas.ac.jp/

Industrial companies active in the Japanese space programme (e.g. Mitsubishi, Kawasaki) publish a range of illustrated promotional material.

The Indian space organization, ISRO, issues a quarterly bulletin *Space India*, though some issues have covered more than quarterly periods. The Department of Space publishes a very complete annual report. ISRO's website is:

> www.isro.org

REPORTS AND BOOKS

Dr Anders Hansson and Dr Steven Maguire: The political economy of indigenous space industries—implications for new entrants. Revised paper, 5th February 1999, of presentation made to IAF Congress, Melbourne, 1998.

Institute of Space & Astronautical Science: *Annual report, 1998*. Tokyo, Ministry of Education, Science & Culture.

ISRO: *Induspace—Indian space programme's partnership with industry*. ISRO, Bangalore, 1988.

ISRO: *PSLV User's manual*. Bangalore.

H. Keneko, H. Itagaki, Y. Takizawa and S. Sasaki: *The Selene project and the following lunar mission*. Paper presented to the 50th conference of the International Astronautical Federation, Amsterdam, 4–8th October 1999.

K. Kasturirangan: Developments in space programmes (unpublished paper).

Junichiro Kawaguchi and Kuinori Tono K. Uesgi: Technology development status of the Muses C sample and return project. Paper presented to the 50th conference of the International Astronautical Federation, Amsterdam, 4–8th October 1999.

Y. Matogawa: *Shusui—Japanese rocket fighter in World War II*. Paper presented to the 50th conference of the International Astronautical Federation, Amsterdam, 4–8th October 1999.

Makoto Nagatomo: SPS activities in Japan in international SPS-related activities (unpublished paper).

National paper of India. 3rd United Nations conference on the exploration and peaceful uses of outer space, Vienna, July 1999.

National paper of Japan. 3rd United Nations conference on the exploration and peaceful uses of outer space, Vienna, July 1999.

S. Sasaki, Y Iijima, K. Tanaka, M. Kato, H. Hashimoto, H. Mitsutani, K. Tsuruda and Y. Takizawa: Scientific research in the Selene mission. Paper presented to the 50th conference of the International Astronautical Federation, Amsterdam, 4–8th October 1999.

T. Tanaka, H. Sasaki and T. Imada: H-II Transfer Vehicle capabilities. Paper presented to the 50th conference of the International Astronautical Federation, Amsterdam, 4–8th October 1999.

BOOKS

Phillip S. Clark: *Soviet rocket engines*. Astro Info Services, 1989.

Kenneth Gatland: *Missiles and rockets*. London, Blandford, 1975.

Kenneth Gatland (Ed.): *The illustrated encyclopedia of space technology*. Salamander, 1981.

Kenneth Gatland (Ed.): *The illustrated encyclopedia of space technology*. 2nd edition, Salamander, 1989.

Gordon Hooper: *The Soviet cosmonaut team*. GRH Publications, 1990.

D.C. King-Hele, D.M.C. Walker, A.N. Winterbottom, J.A. Pilkington, H. Hiller, G.E. Perry: *The RAE table of Earth satellites, 1957–1989*. Royal Aerospace Establishment, 1990.

David M. Harland: *The space shuttle—roles, missions and accomplishments*. John Wiley, 1998.

Nicholas Johnson: *The Soviet reach for the Moon*. Cosmos Book, 1994.

Mohan Sundara Rajan: *Indian in orbit*. Government of India, 1997.

Tom Morgan (Ed.): *Jane's spaceflight directory, 1998-9*. Jane's, 1998.

Reginald Turnill: *The observer's book of unmanned spaceflight*. London, Frederick Warne, 1974.

PERIODICALS

NASDA: *NASDA report*. No 1–. Monthly, Tokyo, NASDA, 1988–.
ISRO: *Space India*. Quarterly, Bangalore, ISRO, 1988–.
Department of Space: *Annual reports*. Annual, from 1988. ISRO, Bangalore.

ARTICLES

Joseph C. Anselmo: Comet, asteroid missions seek clues to Earth's origins. *Aviation Week & Space Technology*, 9th December 1996.
Fabrizio Bernadini: Inside the ISS. *Spaceflight*, vol. 41, January 1999.
Darren Burnham: Japan approves new Moon mission—probes to study lunar surface. *Spaceflight*, vol. 33, July 1991.
Darren Burnham: Return to the Moon: Lunar missions face uncertain futures. *Spaceflight*, vol. 33, November 1991.
Darren Burnham: Hiten mission ends—Buddhist angel's final resting place on Moon. *Spaceflight*, vol. 35, August 1993.
Darren Burnham: Japan's spear of destiny—Moon mission will strive to score lunar bull's eye. *Spaceflight*, vol. 39, August 1997.
Stephen Byford, Clive Simpson and Nik Steggall: Will Japan steal the thunder? Successful launch heralds a new era. *Spaceflight*, vol. 28, November 1986.
Dr A.M. Choudhury: Satellites saving lives. *Spaceflight*, vol. 25, #6, June 1983.
Craig Covault: India launches new booster. *Aviation Week & Space Technology*, 24th October 1994.
Craig Covault: Japan to monitor whales from space. *Aviation Week & Space Technology*, 16th October 1995.
Craig Covault: Japan accelerates station development and promotion. *Aviation Week & Space Technology*, 29th January 1996.
Craig Covault: Japanese satcom companies gain momentum in Pacific. *Aviation Week & Space Technology*, 4th August 1997.
Craig Covault: Japanese H-II failure ruins satcom research mission. *Aviation Week & Space Technology*, 2nd March 1998.
Craig Covault: Advanced satcoms pace Asian space technology; Japanese technology missions readied; Japanese Moon missions advance toward launch; Japanese Mars manoeuvres set. *Aviation Week & Space Technology*, 7th December 1998.
Dr J.K. Davies: MOS-1: A Japanese landsat. *Spaceflight*, vol. 25, #7–8, July/August 1983.
Neil W. Davis: Japan–first with DBS. *Spaceflight*, vol. 25, #5, May 1983.
Neil W. Davis: Japan's new rocket engine. *Spaceflight*, vol. 25, #7–8, July/August 1983.
Neil W. Davis: Japan's Halley's comet probe. *Spaceflight*, vol. 25, #6, June 1985.
Neil W. Davis: Japan's Halley probes. *Spaceflight*, vol. 26, December 1984.
Neil W. Davis: Japanese space science. *Spaceflight*, vol. 27, March 1985.
Dwayne A. Day: Reconnaissance for the rising sun. *Spaceflight*, vol. 41, #10, 1999.
John Elliott: Communications satellite boost for Indian unity. *Irish Times*, 26th September 1983.

Brian J. Ford: The rocket race, in Barrie Pitt (Ed.): *History of the second world war.* London, Purnell, 1969, vol. #6.

Tim Furniss: India aims for polar launcher. *Flight International,* 3–9 June 1992.

AKS Gospalan: 25 years of remote sensing activities in SAC. *SAC Courier,* vol. 23, #1, January 1998.

Gordon Hooper: Year of the journalist. *Spaceflight News,* #53, May 1990.

Sam Jones: Japan's space plans. *Flight International,* 27th September 1973.

Neville Kidger: Japan's new launcher, *Spaceflight,* vol. 23, #8, October 1981.

Neville Kidger: Japanese space processing. *Spaceflight,* vol. 23, #8, October 1981.

Neville Kidger: India's SLV-3 launch vehicle. *Spaceflight,* vol. 24, #2, February 1982.

Neville Kidger: Japanese space plans for 1982–1985. *Spaceflight,* vol. 24, #5, 1982.

Neville Kidger: Salyut mission report. *Spaceflight,* vol. 26, September/October 1984.

Neville Kidger: Japanese cosmonaut reporter candidates. *Zenit,* #34, December 1989.

Neville Kidger: Japanese reporter in space—first fare-paying cosmonaut on Mir. *Spaceflight,* vol. 33, January 1991.

Nichlas D. Kristof: Japan plans spy satellites over a suspicious region. *International Herald Tribune,* 7th February 1999.

Bill Lai: National subsidies in the international commercial launch market. *Space Policy,* vol. 9, #1, February 1993.

Christian Lardier and Vivek Raghuvansji: Le PSLV intéresse les militaires indiens. *Air & Cosmos,* 1491, 28 October 1994.

Christian Lardier: Croissance de 9% du budget NASDA en 1995? *Air & Cosmos.* #1499. 23 December 1994.

Christian Lardier: Lancements et reports de tirs Japonais. *Air & Cosmos,* 1531, 15 September 1995.

Christian Lardier: Les futurs projets japonaise jusque'en 2006. *Air & Cosmos,* 1535, 13 October 1995.

Christian Lardier: Le Japon veut réduire les couts d'ici à 2005. *Air & Cosmos,* #1553, 16 February 1996.

Christian Lardier: Echec de la récuperation d'HYFLEX. *Air & Cosmos,* #1553, 16 February 1996.

Christian Lardier: Maintien du budget de la NASDA en 1996. *Air & Cosmos,* #1569. 7 June 1996.

Christian Lardier: Deux nouveaux petits satellites japonais. *Air & Cosmos.* #1598, 31 January 1997.

Christian Lardier: Premier vol réussi du lanceur Nippon M-5. *Air & Cosmos,* 1600, 14 February 1997.

Christian Lardier: Le Japon prepare de nouveaux satellites. *Air & Cosmo.* #1632, 31 October 1997.

Christian Lardier: Nouveau problème pour le Japon. *Air & Cosmos.* #1632, 31 October 1997.

Christian Lardier: Un budget spatial indien en forte hausse. *Air & Cosmos,* 1610, 25 April 1997.

Christian Lardier: Le budget de l'espace japonais en croissance. *Air & Cosmos,* 1703, 14 May 1999.

Christian Lardier: Le vol de l'HTV japonais prévu en aout 2002. *Air & Cosmos*, 2 July 1999.

H.P. Mama: India's Earth resources satellite. *Spaceflight*, vol. 21, #7 July 1979.

H.P. Mama: Launch vehicle developments. *Spaceflight*, vol. 28, December 1986; and unattributed article: India—the way forward, same issue.

H.P. Mama: India's rocket propellant developments. *Spaceflight*, vol. 37, January 1995.

H.P. Mama: Remote sensing—key to India's future. *Spaceflight*, vol. 38, April 1996.

James T. McKenna: US, Japanese crew hone orbital repertoire. *Aviation Week & Space Technology*, 22nd January 1996.

Michael Mecham: Japan's launch capacity to grow with J-1, *Mu-5*; Japan broadens scope of science missions. *Aviation Week & Space Technology*, 20th March 1995.

Michael Mecham: Indian space—success on a shoestring (series of three articles). *Aviation Week & Space Technology*, 12 August 1996.

Michael Mecham: Japan launches new eye for environmental studies. *Aviation Week & Space Technology*, 26th August 1996.

Michael Mecham: Instruments coming alive on ADEOS-1 Earth scanner. *Aviation Week & Space Technology*, 23rd September 1996.

Michael Mecham: Imagery a vital planning tool; Smallsats open regional door. *Aviation Week & Space Technology*, 7th December 1998.

Michael Mecham: ADEOS 1 loss sets back global environmental studies. *Aviation Week & Space Technology*, 7th July 1997.

Michael Mecham: Stability, science mark Japanese contribution. *Aviation Week & Space Technology*, 8th December 1997.

Michael Mecham: India test-fires *Agni* missile; Pakistan responds with *Gauri*. *Aviation Week & Space Technology*, 19th April 1999.

Michael Mecham and Eiichiro Sekigawa: Japanese reschedule more H-IIA launches. *Aviation Week & Space Technology*, 19th August 1999.

Nilamani Mohanty: Satellite communications activities—a perspective. *SAC Courier*, vol. 23, #1, January 1998.

Theo Pirard: Japan—three new vehicles before the year 2000. *Spaceflight*, vol. 35, October 1993.

Theo Pirard: Toward the Moon and the solar system. *Spaceflight*, vol. 38, August 1996.

Theo Pirard: Japan's very busy launch schedule—technological spacecraft tests in 2000–2001. *Spaceflight*, vol. 39, June 1997.

Theo Pirard: PROBA sera lancé par le PSLV en 2000. *Air & Cosmos*, 1646, 20 February 1998.

Theo Pirard: *Artemis*—next European technological satellite. *Spaceflight*, vol. 40, September 1998.

Joel W. Powell: Long baseline radio telescope. *Spaceflight*, vol. 39, July 1997.

Joel W. Powell: Tropical rainfall research and automated space dockings. *Spaceflight*, vol. 41, #8, August 1999.

Radhakrishna Rao: Remote sensing in India. *Spaceflight*, vol. 25, December 1983.

Andy Salmon: Express delivery. *Spaceflight*, vol. 39, January 1997.

Roloef Schuiling: First Japanese Spacelab. *Spaceflight*, vol. 35, April 1993.

Roloef L. Schuiling: STS-87—microgravity research, satellite rescue and ISS technology tests. *Spaceflight*, vol. 40, April 1998.

Eiichiro Sekigawa: Japan committed to space station. *Aviation Week & Space Technology*, 5th April 1993.

Eiichiro Sekigara: High H-II cost worries NASA. *Aviation Week & Space Technology*, 19th July 1993.

Eiichiro Sekigawa: Renewed backing for Japan's HOPE. *Aviation Week & Space Technology*, 2nd August 1993.

Eiichiro Sekigawa: Japan to stress reusable vehicle. *Aviation Week & Space Technology*, 8th August 1994.

Eiichiro Sekigawa and Michael Mecham: Mitsubishi advances LE 5 design as H-II goes commercial. *Aviation Week & Space Technology*, 15th July 1996.

Eiichiro Sekigawa and Michael Mecham: NASDA plans November launch for ETS 7, TRMM. *Aviation Week & Space Technology*, 14th April 1997.

Eiichiro Sekigawa: Japan forced to cut HOPE, other programmes. *Aviation Week & Space Technology*, 4th August 1997.

Eiichiro Sekigawa: NASDA seeks cause of LE-5A motor casing burn through. *Aviation Week & Space Technology*, 16th March 1998.

Eiichiro Sekigawa: Japanese watchdog agency questions need for J-1. *Aviation Week & Space Technology*, 11th May 1998.

Eiichiro Sekigawa: Budget pressures takes toll on HOPE programme. *Aviation Week & Space Technology*, 13th July 1998.

Eiichiro Sekigawa: Japan considers hybrid satellite defence options. *Aviation Week & Space Technology*, 2nd November 1998.

Eiichiro Sekigawa: Mitsubishi recce plan gains ground in diet. *Aviation Week & Space Technology*, 9th November 1998.

Eiichiro Sekigawa: Japanese space failures linked to lack of coherence. *Aviation Week & Space Technology*, 8th February 1999.

Eiichiro Sekigawa: JEM, Satcom funds edge up. *Aviation Week & Space Technology*, 1st March 1999.

Eiichiro Sekigawa: Japan's H-IIA launcher prompts manifest changes. *Aviation Week & Space Technology*, 14th June 1999.

Clive Simpson: *Tenma*—Japan's x-ray satellite. *Spaceflight*, vol. 26, June 1984.

Nicholas Steggall: Japan's geostationary meteorological satellite system. *Spaceflight*, vol. 24, #6, June 1982.

Bert Vis: Payload specialist flight hopes. *Spaceflight*, vol. 31, April 1989.

Steven Young: Commercial Atlas Centaur fails—$35m Japanese TV satellite lost. *Spaceflight*, vol. 33, June 1991.

References

[1] For example, Ryojiro Akiba: Professor Itokawa's contribution to space research in Japan. Unpublished paper supplied to the author, 1999.

[2] Joseph Treen and David Lewis: Japan's troubles in space. *Newsweek*, 28 May 1983.

[3] Michael Mecham: Japan space programmes keyed to H-II success. *Aviation Week & Space Technology*, 31 January 1994.

[4] NASDA report #24, spring 1994.

[5] Space Activities Promotion Council: *Space in Japan*. Japan Federation of Economic Organizations.

[6] Marco Antonio Caceres: Satcom market buffeted by economic uncertainties. *Aviation Week & Space Technology*, 11 January 1999.

[7] Japan space probes in unmanned docking. *The Times*, 8 July 1998.

[8] NASDA report #7, March 1990.

[9] This is the project description given by Kawasaki.

[10] Astronautics in ancient India, from *Space India*, Oct–Dec 1988.

[11] India's sputnik completes its first six months. *Soviet Weekly*, 1 October 1975.

[12] Comments made during 50th International Astronautical Federation conference, Amsterdam, 4–8th October 1999.

[13] *Space India*, October 1993 to March 1994.

[14] Japan studies mini-shuttle. *Flight International*, 8 March 1986.

[15] Christian Lardier: La propulsion future sera cryogenique. *Air & Cosmos*. #1509, 10 March 1995.

[16] David S. Cloud: Warheadache. *The New Republic*, 20 April 1998.

Index